ALL NEW
KITCHEN
IDEAS THAT WORK

ALL NEW

KITCHEN
IDEAS THAT WORK

HEATHER J. PAPER

The Taunton Press

DEDICATION

To Russ

The Taunton Press
Inspiration for hands-on living®

The Taunton Press, Inc.
63 South Main Street, PO Box 5506
Newtown, CT 06470-5506
e-mail: tp@taunton.com

Editor: Peter Chapman
Copy Editor: Nina Rynd Whitnah
Indexer: Heidi Blough
Interior Design: Carol Petro
Layout: Jodie Delohery
Illustrators: Joanne Kellar Bouknight p. 69; Barbara Cottingham pp.148–149
Cover Photographers: Front cover photos (top, left to right): Tria Giovan, Andrea Rugg Photography/Collinstock, Mark Lohman, Rob Karosis Photography/Collinstock; (large photo): Helen Norman
Back cover photos (clockwise from top left): Chipper Hatter, Mark Lohman, Helen Norman, Helen Norman, Ryann Ford, Helen Norman

The following names/manufacturers appearing in *All New Kitchen Ideas that Work* are trademarks: Bluetooth®, Caesarstone®, Cambria®, Corian®, EcoTop®, Energy Star®, Formica®, Silestone®, Wilsonart®, Zodiaq®

Library of Congress Cataloging-in-Publication Data

Names: Paper, Heather J., author.
Title: All new kitchen ideas that work / Heather J. Paper.
Description: Newtown, CT : Taunton Press, [2018] | Includes bibliographical references and index.
Identifiers: LCCN 2018003416 | ISBN 9781631869013
Subjects: LCSH: Kitchens–Remodeling. | Kitchens–Design and construction.
Classification: LCC TH4816.3.K58 P37 2018 | DDC 690/.44–dc23
LC record available at https://lccn.loc.gov/2018003416

Printed in the United States of America
10 9 8 7 6 5 4 3 2 1

ACKNOWLEDGMENTS

Giving an author full credit for a book is no different than singling out an actor for the triumph of a movie. Just as a movie's success also depends on writers, producers, directors, and photographers—just to name a few—so does a book depend on the talents of many.

First and foremost, I have to thank Peter Chapman, Executive Editor at The Taunton Press. I've had the pleasure of working with Peter on several projects and appreciate the confidence he's always had in me. He seems to think that I can tackle almost any subject and for that alone I'm grateful; any writer will tell you that it's impossible to have too much self-assurance.

My sincere thanks go, as well, to Rosalind Loeb, one very talented—and organized—art director. As you can tell from the vast number of photographs in this book, Rosalind plays an important role, accounting for every image and perfecting its place on each page.

Credit also goes to numerous others, including the homeowners and design professionals represented in this publication. I never tire of looking at a photograph, wondering how a homeowner, kitchen designer, interior designer, or architect came up with a concept and then had it translated into reality by builders, contractors, and craftsmen. Thank you all for generously sharing your work with us.

A sincere thank you goes, too, to the photographers who provided top-notch images for this book. Mark Lohman, Chipper Hatter, Hulya Kolabas, and Helen Norman all invested time in locating, and shooting, extraordinary kitchens. But kudos go to others, as well; please reference the credits at the back of the book for the names of photographers—and design professionals—for individual images.

Finally, I want to thank my supportive family, friends, and mentors, past and present. Most of all, though, I thank my husband, Russ, whose unwavering love and encouragement mean more than he will ever know.

—Heather J. Paper

CONTENTS

INTRODUCTION

From the day I moved into my first apartment, I've been honing my idea of the "perfect kitchen." Granted, it wouldn't take much to improve upon that 5-ft. by 8-ft. galley space, but designing any dream kitchen can be overwhelming.

On the pages of this book, we'll take you through the process step by step. You probably have an idea in your mind's eye what you want the finished space to look like. But it's important, first, to take a step back—several steps, for that matter. Are you planning to simply make some cosmetic changes? Remodel your existing kitchen? Or go all out, knocking down walls to make room for a new addition? Only after you've made that decision can you get down to some careful space planning. Which elements of your current kitchen will stay—if any—and which will go? How do you see today's latest technology fitting into your overall plan? And, from an aesthetic point of view, how will your kitchen blend with adjacent spaces?

As you shop for the new elements of your dream kitchen, be sure to consider every alternative. You can do it without ever leaving the house; shopping online is one of the best ways to compare products. When it comes to making final decisions, however, be sure to check out your choices in person. (An online photo doesn't always accurately represent how an item looks.) Keep an eye, too, on seasonal sales that can help you stretch your budget.

Whether you're self-confident or a bit uncertain when it comes to kitchen design, you'll want to rely on expert opinions along the way. Kitchen designers, interior designers, and architects can be invaluable for cutting-edge concepts, while builders, contractors, and craftsmen can bring their ideas to life. Every step of the way, though, be sure that their visions are in line with your own sense of style. Even the most efficient, hard-working kitchen won't be welcoming if it doesn't reflect your personal preferences.

Filled with information and inspiration, *All New Kitchen Ideas that Work* is sure to conjure up ideas for your own dream kitchen; I know it has mine. I'm considering upgraded cabinetry and countertops, appliances that can sync with my smart phone, even seating that will encourage friends and family to hang out while I'm cooking. But those are my priorities. Whatever yours may be, you'll find thought-provoking views throughout this book. Here's to a kitchen beyond your wildest dreams!

THE BIG

Whether planning a facelift or a full-scale remodel,

turning the kitchen of your dreams into a reality—without breaking the bank—

will require some important up-front decisions.

PICTURE

Assess Your Needs

The logical place to start any kitchen project is by assessing your needs. You may think that's just a matter of deciding which elements you like and which ones you don't. And that's certainly part of the process. But it's also essential to step back further, to take a look at the big picture. Is your kitchen part of an open floor plan, one that includes a dining and/or living area? If so, existing architectural elements such as fireplaces, for instance, should coordinate in terms of style. Even if your kitchen is a separate space, adjacent rooms that are clearly visible should be taken into account; cabinetry finishes, for example, should not only complement the kitchen floor but also the floors of rooms in plain view.

Next, practical matters come into play; your kitchen should function to suit your lifestyle. If you're the lone cook, a single sink and stove should do, but if you and your spouse like to work in tandem, consider two ovens and a six-burner, restaurant-style stove. Or, if yours is a busy, always-on-the-go family, you might opt for a microwave located just outside the primary work area, convenient for grab-and-go meals.

More people today are opting to age in place, too, planning for the future with universal design. This increasingly popular concept goes well beyond ramped entryways and wheelchair-width passageways. Even the simplest details, like opting for wing-style faucet handles that don't require a turn of the wrist, can make a difference in your kitchen's long-term usefulness. At its best, universal design accommodates every family member at each stage of life.

Finally, take a hard look at what already works well and which elements you want to keep; these will often be the jumping-off points for your kitchen's design. You may, for instance, have a slate floor that you can't bear to part with, so you'll want to choose cabinetry and countertops to coordinate with it.

One by one, assess your kitchen's layout, appliances, storage, counter space, flooring, fixtures, and lighting. Proceeding in this kind of step-by-step manner will help you decide if a face-lift, a renovation, or a remodel is right for you.

When the kitchen is part of a larger great room, it's not always desirable to see the appliances. They all but disappear in this low-key kitchen; even the range hood has a sleek, streamlined look.

Taking advantage of this kitchen's stunning view, storage along two walls is limited to base cabinets, with windows reaching from the countertop to the ceiling. There's no shortage of storage, however; a tall unit on an adjacent wall makes up for the absence of wall cabinets.

Take Your Time

If you've recently moved into a new residence and want to remodel the kitchen, take some time to complete your assessment. Using your kitchen for a few months will reveal its strengths and weaknesses, allowing you to invest your hard-earned dollars where they'll do the most good. When you finally start the remodeling process, you may be surprised to find that an original wish-list item no longer has a place in your overall plan.

above • For a family who spends a lot of time in the kitchen, a spacious island with counter stools makes sense. It's the perfect place for dining any time of the day and it provides a larger work surface for the cook.

left • In an otherwise streamlined kitchen, a rustic barn door leading to the pantry adds a touch of personal style. At the same time, it's easy to roll open and shut—always a plus for the busy cook.

left • Although all-white cabinetry is a can't-go-wrong style, the owners of this home elevated the look with select black and stainless-steel elements. Cabinet hardware and perimeter countertops bring the black into play, while contemporary stools combine the two tones as does the flared range hood—the focal point of the room.

left • Proof that kitchens need not have a sterile feeling, this one has the same country appeal as the rest of the house. Cream-colored cabinetry, comfortable seating, and soft window shades add to the appeal. But it's the rustic beamed ceiling that completes the look.

right • By keeping the sink and appliances around this kitchen's perimeter, there's no need to build anything into the island. That frees up the entire surface for food prep, casual dining— and even spreading out homework.

left • If two cooks typically use the kitchen, two islands can keep one from getting in the way of the other. Each of these islands has its own sink and storage space but that's where the similarities end; one is reserved for food prep while the other doubles as a breakfast bar.

Count on Soothing Neutrals

To create a sense of cohesiveness between your kitchen and rooms that open onto it, go with the flow; introduce elements that are common to all spaces. You might carry through a color or two, echo architectural details, or simply repeat decorative motifs.

This kitchen relies on neutral hues to connect it with the adjacent living and dining areas, with shades of white starring in all three spaces. In the kitchen, crisp white is splashed across cabinetry as well as the sink and backsplash, then repeated in the living room's sofa and the dining area's upholstered chairs and shiplap wall covering. To keep the collective rooms from seeming too stark, however, soft grays and warm woods enter the mix, showing up in marble countertops, upholstered furniture, and barstools that have a sculptural quality.

Even finishing touches provide a sense of unity to the three spaces. Rich metallics make an appearance in faucets and fittings, light fixtures, cabinetry hardware—even the bases of tables and barstools—resulting in rooms that are right at home with one another.

above • This white-on-white kitchen is elegant in its own right, further enhanced by the choice of light fixtures—a brass pendant over the sink and a beaded fixture over the island. Both would be equally at home in the adjacent living area.

right • The brass faucet, light fixture, and cabinet hardware are stylish and refined, making them look more like fine jewels than functional necessities. For a sense of continuity, the metallic finish appears in the living area, too.

above · Taking advantage of every square inch, drawer storage is built right into the kitchen island. Opposite the sink wall, to-the-ceiling storage—with the same luxurious white finish that's found in the living and dining areas—includes an integrated refrigerator/ freezer and pantry.

left · Given the kitchen's elegant materials and finishes, the fine furnishings in this dining area are a good fit. Another thing the two spaces have in common is exquisite light fixtures; the chandelier beautifully illuminates the table below.

A Facelift vs. a Full Remodel

Your kitchen may not need a full remodel; sometimes a simple facelift will do. (That may be all your budget will allow, too.) Your needs assessment will help determine which option is right for you, but what's the difference between the two options?

If your current cabinets are structurally sound and you're keeping the existing footprint to preserve the flooring, your kitchen could be an excellent candidate for a facelift. While this is a less-expensive option than remodeling or renovation, a facelift won't address major problems, such as not enough light or lack of space. On the other hand, a facelift can make your kitchen more efficient and give it a fresh appearance. Refacing or refinishing cabinets, for instance, won't give you more storage but will give you a new look. And you can make your cabinets more efficient by retrofitting them with storage accessories. Or, you might opt to bypass your cabinets entirely, resurfacing your countertops instead. Most often, this change will be from laminate to engineered stone. Keep in mind, however, that this type of change usually requires replacing the faucet, sink, and drains, too. On the upside, it also provides the opportunity to upgrade to a sink with a better bowl configuration and perhaps a pull-out faucet.

A remodel, on the other hand, involves major changes that may take your kitchen in a whole new design direction. Maybe you're ready to turn a tired traditional kitchen into a stunning contemporary space. That might require changing the kitchen's footprint to add space or, at least, reshape the room. From there it's a matter of incorporating everything from stunning cabinetry and smart appliances to brand-new windows and skylights, even breakfast nooks.

Whether you opt for a facelift or a full-scale remodel, keep a firm grasp on the bottom line. Home projects often end up costing more than expected, but careful budgeting along the way will help keep expenses in check.

If your cabinets are in good shape, perhaps a fresh coat of paint is all they need. After cleaning, repairing nicks and dings, sanding, and priming, paint them the color of your choice. New hardware can make a big difference, too.

Resurfacing Countertops

Resurfacing systems now allow stone tops to be directly installed over existing counters—an especially attractive option for laminate that's worse for the wear. These countertop systems are most often made from a thin layer of engineered stone, delivering good looks and top-notch performance much like a standard 2cm or 3cm slab of stone. Concrete and porcelain resurfacing systems are also available, as are those for granite and marble tile. Several come in kit form, too, providing do-it-yourself options.

The primary benefits of these resurfacing systems are reduced demolition (adding less to landfills) and not as much kitchen down time. The more sophisticated systems, such as engineered stone, should be left to professional installers. You can find those in your area by looking up "countertop resurfacing" on the Internet, but be sure to check references and local licensing requirements.

above • Even a kitchen with basic white cabinets can be given a fresh appearance. You might, for instance, give an island a new look by painting the base or facing it with a new wood or material. This kitchen takes the concept a step further; the wood used on the base of the island reappears as ceiling beams.

left • To keep your budget in check, keep a close eye on your priorities. In order to have the appliances of your dreams, for instance, you might forego brand-new cabinets and opt to reface or refinish your current ones instead.

If you're torn between traditional and contemporary styles, a transitional kitchen may be for you as it blends elements of both. Here, a coffered ceiling, mullioned windows, and an elaborate vent cover represent traditional elements, whereas the streamlined cabinetry—especially the tapered legs—are more contemporary.

What's Your Style?

Maybe you have a clear vision of your dream kitchen. If not, take a look at your current space; which elements do you want to keep or replicate in a remodel? Which ones no longer appeal to you? Settling on your kitchen's style has much to do with your home's architecture, too, particularly if it opens up to adjacent rooms.

Styles that withstand the test of time include traditional, transitional, country, and contemporary. According to the National Kitchen & Bath Association (NKBA), transitional kitchens are currently the most popular with contemporary and traditional coming in second and third. Country style continues to have a fan base, too, while the popularity of industrial and mid-century modern styles is on the rise. What's in fashion at the moment, though, should always take a back seat to your family's lifestyle and your personal preferences.

Collect Your Thoughts

If you're undecided about your personal style, look to the multiple sources available today for design inspiration. Thumb through magazines—and this book. Tour home shows in your area. Go to websites, blogs, social media pages, even smart phone apps. Look for rooms that you find appealing; there are sure to be similarities, giving you a sense of where to start. Look for qualities that identify a certain style, too, such as the warmth of a country-style kitchen or the sleek lines of a contemporary space. Don't be surprised if you find that you like a variety of styles; an eclectic approach can be one of the best ways to put your personal stamp on a room.

The key is to stay organized. The more ideas you collect, the easier it will be for your professional team to create the kitchen you envision.

above • Marble countertops and face-frame cabinets—some of them glass-fronted— give this kitchen an overall traditional feeling. At the same time, though, there's a contemporary twist, expressed in the stainless-steel hood and pendant lights.

left • Although wood walls, floors, and ceilings are often associated with traditional style, those elements have been handled here in a contemporary way. Walls are painted white to match the sleek kitchen cabinets, while the floor and ceiling are natural ash, a stunning contrast to the cabinetry.

TRADITIONAL

Traditional kitchens come in many incarnations, but they all have one thing in common: They are based on the architecture of bygone eras, typically before the mid-20th century. As a rule, traditional kitchens have more ornate moldings than their modern counterparts, while cabinetry often features frame-and-panel doors. Certain styles do call for specific details, such as inset quarter-sawn oak doors on Craftsman-style cabinets, but don't assume that all traditional kitchens are brown. White kitchens are a perennial favorite, while gray schemes now come in a close second. Two-toned kitchens are gaining in popularity, too; cream-colored perimeter cabinets might be teamed with a soft gray island, or a kitchen with espresso-stained cabinetry might feature a pale wood floor. No matter how you mix things up, the result will be your own fresh twist on tradition.

above • In the opinion of many kitchen experts, gray is the new white when it comes to kitchens. Here, the two neutrals marry beautifully; a gray tone—similar to that found in the floor—washes opposing walls as well as the island between them. Crisp white cabinetry, meanwhile, provides the perfect contrast.

above • Ornate detailing is the hallmark of traditional kitchens, represented here not only by the cabinetry and ceiling but also by the elaborate vent cover. Behind the range, an intricate marble tile backsplash befits the grand architectural element.

left • Rich wood cabinetry and stone countertops often characterize traditional design. But the stone aspect is taken a step further here; the material is also used to create a luxurious backsplash as well as an impressive hood over the range.

CONTEMPORARY

While a modern kitchen is contemporary in style and a contemporary kitchen can be modern, the terms aren't necessarily interchangeable. Modern style refers to a look rooted in the early- to mid-20th century, whereas contemporary is a much broader term. Contemporary kitchens are identified by their streamlined style, with smooth surfaces, recessed lighting, and full flush doors in frameless cabinets, typically with simple drawers and doors. Gleaming materials—such as stainless steel, glass tile, and high-gloss composites—are most often the finishes found in a contemporary kitchen, but even traditional materials can take on a contemporary point of view if designed in a clean, linear way. It's not the specific material so much as the understated detailing that's the hallmark of a contemporary kitchen.

above • A mix of materials is represented in this kitchen, each one as clean-lined and streamlined as the next. Laminate cabinetry and quartz countertops are the perfect partners for built-in appliances that all but disappear.

left • Stainless steel is a material often found in contemporary kitchens—in appliances, sinks, and even hardware. But this kitchen uses stainless steel in a more unusual way; it's incorporated into one end of the island.

Tile and wood paneling in nearly the same gray hue create a subdued backdrop in this contemporary kitchen, allowing unique elements to take center stage. Brilliant blue stools are eye-catching against the neutral backdrop, while a multifaceted black light fixture illuminates the island.

The perimeter of this kitchen is perfectly traditional, with white face-frame cabinets lining the walls. But the center of the space is where more contemporary elements come into play. Although the island utilizes the same cabinetry, it's painted gray, complemented by rattan counter stools and globe-shaped pendants.

TRANSITIONAL

Combining traditional and contemporary styles, transitional design has a clean, classic look. If you admire the finer details of traditional rooms but also appreciate sleek contemporary spaces, a transitional kitchen will give you the best of both worlds. Find common ground in the elements of your room, so the disparate styles complement— and don't compete—with one another. One of the easiest ways to do that is with color; spread neutral hues throughout the space, for instance, in the form of cream-colored cabinets, a coffee-colored tile backsplash, and a hardwood floor. Then incorporate stainless-steel accents or minimalist brass hardware.

Some styles, too, are transitional by their very nature. Shaker, for instance, is a prime example; it's timeless with lines so clean that they come off as contemporary.

Although wood is often associated with traditional kitchens, it's all in the way you use it—a point proven in this kitchen. Contemporary white cabinetry is complemented by a T-shape island and range hood that show off their wood grain in an equally modern way. Still, wood beams throughout the space bring it back to a transitional level.

At first glance, the clean lines of this kitchen identify it as a contemporary space. Upon closer inspection, however, it's truly transitional, thanks to a quartet of bentwood counter stools—representing classic traditional design.

COUNTRY

Their interpretations may be wide-ranging, but all country-style kitchens have a look that's instantly warm and welcoming, harkening back to an earlier era in some way. French Country, English Country, and Tuscan style were inspired by rural life in their respective geographic areas. And American Country has a wide variety of incarnations. It's often characterized by muted colors and worn finishes—from reclaimed cabinetry and beamed ceilings to slate floors and butcher-block countertops. Or you might opt for soft colors and feminine details to create cottage style. It's hard to go wrong with crisp white finishes, but pale blues and yellows also have a fresh feeling; beadboard wall coverings and farmhouse sinks can further enhance the look. However you prefer to translate country style, it's sure to be appealing.

With all-white cabinetry and a backsplash to match, simplicity is the appeal of this country-style kitchen. A wood-topped island is a welcome reprieve from the large expanse of white, as are the wooden counter stools. Finally, though, metal light fixtures define this space as unmistakably country.

left • Pine is used in many a country kitchen; here it's given star status in the form of an intricately carved island. The oversize surface is convenient for food prep and close enough to the range and refrigerator to provide a landing area for both.

facing page • Everything about this kitchen says "country," albeit in a sophisticated way. Cream-colored cabinets are complemented by a wide-plank wood floor, beams on the ceiling, and even Windsor chairs around the table. But the details make a difference, too; pewter accessories and gingham accents are the epitome of country style.

Farmhouse Style

Farmhouse-style kitchens recall a time when rural living—and its simplicity—were the norm. Because that kind of straightforward approach is still highly regarded, farmhouse style continues to show up in many a kitchen. Some of its most common elements include the following:

- **Open shelving.** In a classic farmhouse kitchen, wall-mounted shelves keep dinnerware in plain sight—easy to grab in an instant.

- **A vintage-style range.** Today's manufacturers know that the stove is often the heart of a farmhouse kitchen. As a result, several offer vintage-style units, right down to the quintessential stovepipe.

- **Classic flooring.** Wood floors are typically found in a farmhouse kitchen, either with a natural finish or painted.

- **A fireplace.** In farmhouse kitchens of old, the hearth formed a central gathering spot. In lieu of a conventional fireplace, a modern-day interpretation might be a wood-burning pizza oven.

- **A large table.** The precursor of the modern-day island, a farmhouse table can pull double duty as a food-prep surface and eating area.

- **An apron-front sink.** With its broad face, deep bowl, and sturdy construction (it's often made of cast iron), an apron-front sink stands up to daily, heavy-duty use.

- **Freestanding cabinets.** Kitchens of the past typically included hutches, pie safes, and other stand-alone pieces. Those freestanding pieces work in today's kitchens, as well, or replicate the look with a built-in unit that has furniture-style details.

above • Made of soapstone, this apron-front double sink is just as good-looking as it is hard-working. What's more, the vintage-style brass faucets befit the farmhouse look.

left • The top shelf in this kitchen is set atop the windows, resulting in a height that's just right for lesser-used items or pieces that are purely for display.

above · This kitchen has all the appeal of farmhouse style but incorporates some modern conveniences, too. Storage is provided by a combination of sage green cabinetry and wall shelves, the latter displaying a collection of yellowware bowls—also true to the farmhouse style.

left · In the center of this space, an island offers both open and closed storage. The butcher-block top extends over both ends of the island, creating a larger work space as well as room to tuck stools under the overhangs.

Develop a Budget

How much you spend on your kitchen will be based on a number of factors. If you're building a new home or embarking on a whole-house remodel, the kitchen will be budgeted as part of your overall per-square-foot spending allowance. Otherwise, your project dollars will be dedicated primarily to your kitchen. Still, even in these scenarios, it's not unusual for a portion of the budget to extend to a neighboring room. New kitchen flooring might extend into an adjacent room, or your choice of backsplash tile might be used on a nearby fireplace surround, too.

When developing a plan, keep in mind that cabinets and appliances can consume a large part of your budget; if required for your door style, cabinetry hardware will need to be budgeted in, too. Countertops can also be a sizable purchase, especially if you're considering engineered or natural stone, while sinks and faucets vary widely in price, based on both their quality and their features. Flooring and lighting costs can add up quickly, as well, especially if you plan to expand them beyond the kitchen into adjacent rooms. Finally, don't forget about incidentals such as drains, soap dispensers, and garbage disposals. Although each is a small item, their combined impact on your budget won't be.

Labor costs will greatly impact your budget, especially if you're relocating or adding plumbing or electrical components, or tearing down walls. (Surprises lurking behind your walls or beneath your floors can affect even the most well-thought-out budget; a good rule of thumb is to factor in an extra 20% for the unexpected, especially if you're remodeling an older home.) Other fees that can add up quickly include everything from design and engineering to permits and even loan charges. Be sure to give some thought as to how you'll handle meals during the remodeling process, as well. Not having access to your kitchen can hit your household budget hard, even spilling over into your remodeling dollars. If you don't plan to use an outdoor kitchen or make do in another part of the house, be sure that your budget includes a line item for restaurant meals and convenience foods.

Custom cabinetry can quickly add to your remodel's bottom line. On the other hand, you can be creative with stock cabinets. A vintage-looking wooden cupboard like the one in this kitchen could be made up of stock base and wall components.

above • It's not unusual for a portion of your kitchen budget to extend to a neighboring room, especially in an open floor plan like this. Here, the hardwood floor runs right into the hallway, so it needs to be accounted for accordingly.

left • Built-in appliances can drive up the price of a remodeling project. A built-in cooktop and wall oven, for instance, are more expensive than a single slide-in range. Likewise, a built-in microwave will cost more than a freestanding model that you can simply set on a shelf.

Assemble a Team

Which professionals you use for your kitchen remodel will be determined by how much you want to have done. For structural work, you'll need a licensed architect or a design/build firm. And you might call on skilled tradespeople for certain components, such as custom cabinets. The specific expertise of a kitchen designer can also be an advantage as can that of an interior designer with a lot of kitchen design experience.

ARCHITECT

If your new kitchen will be part of a home addition, architects can be your best assets; these professionals can help achieve harmony between the new and old parts of the house. As needed, they will bring in kitchen professionals to provide cabinetry, appliances, and other selections involved, as well as the builder or general contractor—if not on staff—to execute the design.

DESIGN/BUILD FIRM

If you are building from the ground up, start by hiring an architect or a design/build firm. (Large-scale kitchen remodels involving older homes or major structural changes sometimes use design/build firms, too.) These companies generally have a builder or general contractor on staff and in-house design services. The latter might be an architect, interior designer, kitchen designer, or any combination of the three.

The fine details of an Arts and Crafts kitchen like this require an assist from an architect or a design/build firm—especially for the custom island and the impressive light fixture over it.

Professionals are often required for a kitchen remodel—or even a facelift, for that matter. In this kitchen, for instance, suspended shelves on either side of the range might require the advice of a professional to know whether they can support the weight of heavy dinnerware.

KITCHEN DESIGNER

If you're planning a new kitchen within the current footprint of your home, call a kitchen designer first. Not only do they have access to the latest planning tools and technology, these professionals have the inside scoop on trends, new materials, and building codes. Their expertise can save you a lot of time, money, and frustration. Once the plan is in place, a general contractor may be called in to finalize the budget and execute the plan.

INTERIOR DESIGNER

If you're planning a facelift instead of a remodel, an interior designer can help bring a fresh look to a room. Like kitchen designers, they have the inside track on the latest trends and materials, and can provide guidance as to which upgrades are worth your money and which ones are not.

OTHER PROFESSIONALS

Maybe a specialized tradesman—a tile installer, painter, plumber, carpenter, or electrician—is all that's needed to accomplish the changes you want. Just keep in mind that any work affecting the home's plumbing, electrical, or ventilation systems needs to be handled by a licensed professional.

If you're planning a new kitchen within the current footprint of your home, a kitchen designer may be your professional of choice. Refacing cabinetry, replacing appliances, adding new light fixtures— they're all within their realm.

THE RIGHT

Before rushing to buy appliances that incorporate the latest technology
or up-to-the minute cabinets and countertops, take time to determine
the layout of your kitchen. Well-planned workspaces and storage will result
in a kitchen that's both fashionable and functional.

LAYOUT

Consider the Footprint

The space-planning phase is the most important part of any kitchen remodel. A kitchen may be filled with stunning materials and finishes, but if the layout doesn't work well, it will all be for naught.

A well-thought-out layout starts with your family's lifestyle. How many cooks use the kitchen and when? Do you want—or have the space for—an eat-in kitchen? Or is yours a busy, always-on-the-go family and casual countertop dining at the island will suffice?

Likewise, appliances come into play. Where can ovens and dishwashers be conveniently placed, without their doors—when open—getting in the way of passersby? Or, if you like to entertain, will one refrigerator be sufficient or would two be better? Give thought, as well, to where the pantry will be located. Will it be within the confines of the kitchen proper or will you have a separate space?

After evaluating these various aspects, the kitchen of your dreams will start to take shape. It may be a galley kitchen, a U-shape kitchen, or an L-shape kitchen with an island. But no matter what configuration you come up with, you'll quickly see whether your new kitchen can fit within the current footprint or if you need to completely change up the space.

above • **If you entertain on a regular basis, be sure to plan your kitchen accordingly. This kitchen, for instance, has generous walkways around the island, allowing guests to move freely. Plus, the counter stools can be pulled away, quickly turning the island into a buffet space.**

Efficiency is always a priority, and that starts with the placement of appliances. It doesn't get much more efficient than this kitchen; the distance from the sink to the refrigerator and the sink to the wall ovens is approximately the same—with the cooktop conveniently between them.

Counter stools needn't line up along one side of the island or even be placed at a 90° angle. If, for instance, your kids tend to do their homework while you're cooking dinner—and your spouse likes to catch up on the day's events at the same time—consider two separate seating areas.

MAKE THE EXISTING PLAN WORK

The size of your kitchen will dictate, to a large degree, whether you can work within the current footprint. Is it still large enough to accommodate your needs or has a growing family outdated it? Just as important, can you make do with the existing layout of windows and walls, cabinets and appliances, even entryways? Or are you desperately short, for example, on storage space?

Retaining the footprint doesn't mean that your kitchen's style and functionality can't be improved. Fresh windows (minus the muntins), new tile—on the walls or the floor—and eye-catching light fixtures can update the style of a kitchen without expensive footprint changes. Similarly, specialty cabinets like pull-out trash bins or wide drawer banks are upgrades you can make. You can replace an existing range with one that has a warming drawer and swap a standard microwave for a model with convection, giving you four appliances in the same footprint as two. Or, if space allows, you might add a peninsula to an L-shape kitchen, providing more cabinet and counter space.

Through it all, however, keep a watchful eye on ventilation and plumbing, as changes to either require professional tradesmen and can entail roof, wall, and floor restructuring. By keeping them in their original positions, you'll be able to stretch your remodeling budget much further.

Replacing a single appliance can give a kitchen an entirely new look. This deep blue commercial-style range, for instance, becomes an instant focal point. Still working within the original footprint, though, is a creative touch; on each side of the range, an antique barrister bookcase provides handy storage.

Phase in Upgrades

If there's no room in your budget to add all of your desired upgrades at once, phase them in gradually. If additional storage is needed, for instance, there are a couple of options that allow you to work within your current footprint. First, maximize the cabinets you already have with interior accessories. Make a blind corner cabinet more efficient with a swing-out accessory that makes the back section usable. Use a two-tier organizer to double drawer capacity. Replace a builder-grade base cabinet half-shelf with a full-depth rollout tray.

Second, look for untapped storage opportunities: Add a rail-based organizing system to your backsplash or hang a pot rack from the ceiling. The possibilities are all around you.

above · If your kitchen currently has only base and wall cabinets—and space allows—consider placing a table in the center of the space approximately the same size as a small island. This pine example serves as a hard-working prep area and, with aluminum stools, a casual eating area, too.

right · Why settle for a standard range hood when you can have one that serves as a focal point as well? Although this one is made of traditional metal, it stands out prominently against the three-dimensional backsplash.

CHANGE UP YOUR DESIGN

Before embarking on a remodel that will alter the room's footprint, it's important to ask yourself a few questions. Will it greatly improve your kitchen's functionality? Will it significantly increase your enjoyment of the space? Will it make your kitchen more efficient? If you can answer "yes" to questions like these, you may be a good candidate for changing up your space.

Not all remodeling projects that involve footprint changes need to be on a large scale. Smaller projects typically leave the existing walls and doorways in place, but might add prep sinks, move appliances, or open a pass through. Larger-scale projects, on the other hand, involve considerably more time, skilled workers—and money. There's also the inconvenience factor. You won't be able to use your kitchen while electrical, plumbing, ventilation, flooring, and structural aspects are being worked on. Structural changes may uncover costly surprises, too, if there are problems beneath your walls or floor that you didn't know existed.

Changing a kitchen's footprint can often be handled with the help of a kitchen designer or design/build pro (see pp. 28–29). But when a footprint change is tied into a room addition—where your remodeled kitchen will expand into a new screened porch, for instance—it's best to work with an architect and engineer, as roof lines and load-bearing walls will be impacted. Retaining their services will be well worth it in the end; you'll have a kitchen and an addition that maintain your home's design and its structural integrity.

Not all remodeling projects that involve footprint changes need to be on a large scale. You might, for instance, opt to add a prep sink in an island. In this kitchen, the countertop is extended, too, creating an eating bar.

In a kitchen used by two cooks at one time, extending a kitchen by a few feet can make room for two islands. Here, the island nearest the ovens is reserved for the cooktop, while the other is fitted with a prep sink, with room to spare for an eating bar.

If you're in need of something more structured than at-the-counter eating but less than a formal dining room, bumping out a bay can give you the space needed for a breakfast nook—the perfect place for meals any time of day.

The Open Floor Plan

There was a time when closed-off kitchens were the norm. But the walls, to a large degree, have come tumbling down; today, open floor plans are often the go-to kitchen style, where home-owners can prepare dinner, dine in comfort, help kids with homework, and even entertain— all in the same space.

Because an open floor plan typically includes the kitchen as well as living and dining areas, certain considerations should go into your scheme.

- As you move from space to space, which ones should be adjacent to each other? It makes perfect sense for the dining table to be placed near the kitchen, but do you then want a seating area nearby? Or perhaps a small workspace?

- Don't let traffic patterns be an afterthought. Think about how people will move from one zone to the next, keeping doorways and stairways in mind, too. You will need paths at least 3 ft. wide to travel through the room comfortably.

- Traffic flow needs to be routed around, not through, the cooking zone. Refrigeration and pantry zones, on the other hand, should be accessible to the chef as well as other members of the household and guests.

- The space between a kitchen island or peninsula and a dining room table should be at least 4 ft., allowing room to walk between the two and diners still to have room to pull out their chairs.

- To soften the look of hard-edged appliances, many homeowners opt for integrated appliances. Featuring cabinetry fronts, with ventilation tucked behind matching millwork, they more quietly blend with furnishings in the rest of the room.

Just a few steps beyond the great room's kitchen area is the pantry. Fitted with wall ovens, the refrigerator/freezer, and plenty of storage, it keeps all the necessities conveniently close yet out of sight.

What makes this open plan work, in part, is that the kitchen is so understated. Only the cooktop and sinks are in this space; the remaining appliances are in the adjacent pantry. In addition, the island looks more like a piece of furniture than a functional element.

An eating area is an inherent part of any open floor plan. Because two sides of this table are served by bench seating, this dining spot makes the most efficient use of space.

The open floor plan allows the cook to keep an eye on the news—or a favorite show—while making dinner. And, a pair of upholstered chairs in front of the fireplace is the perfect place for one-on-one conversation any time of day.

Universal Design

While many believe universal design is aimed solely at people who have special needs, the concept allows people of all ages and physical abilities to live comfortably and safely. Here's how universal design principles can be adapted to some basic areas in the kitchen.

Multiple countertop heights: Because most kitchens accommodate more than one person, include multiple countertop heights that will provide workspaces for everyone. To accommodate someone who's seated while taking on kitchen tasks, incorporate at least one countertop that's 30 in. or less. (A standard table can also be used.) For tasks that require standing, the best countertop height is 3 in. shorter than your elbow height.

Seating: If you have an island or a peninsula that accommodates seating, make the counter 30 in. high—the same as table height. This offers optimal seating for everyone, regardless of age or mobility. Young and old alike can use standard chairs, and a wheelchair can pull up to the counter, too.

Smart appliance planning: Wall ovens and microwaves mounted at shoulder height are easier to reach into than standard ovens; you can reach straight in and see what you're pulling out, which is a safer strategy, too. Likewise, dishwasher drawers are an ergonomic alternative to the traditional dishwasher with a pull-down door. Two dishwasher drawers positioned on either side of the sink will match the capacity of a single, standard model.

Easy-care surfaces: Because universal design is about making the kitchen easy to use in every way, consider low-maintenance countertop surfaces. Budget-friendly laminate or solid surfaces such as Corian® are easy to clean and durable. A side benefit of the latter is that they can accommodate undermount sinks; because undermounts have no lip around the perimeter, food crumbs can be wiped right into the sink.

Comfortable, slip-resistant flooring: Standing on hard surfaces for long periods can take a toll on anyone, so use cork, wood—even cushioned mats—instead of hard, slick surfaces. If your heart's set on stone flooring, however, consider coating it with an anti-slip treatment.

Cabinetry: Well-thought-out cabinets are all in the details. In terms of hardware, opt for pulls with plenty of finger space (such as D-rings), touch latches, or motion openers. Roll-out trays, lazy Susans, swing-out storage, and pull-down shelves make cabinet interiors more easily accessible, too.

Traffic paths: A kitchen based on universal design needs to have ample space to move around. Plan traffic paths that measure between 42 in. and 48 in., allowing more than one cook in the kitchen and anyone in a wheelchair to pass easily.

Not only is this desk dropped down to a height that meets Americans with Disabilities Act (ADA) standards, the 4-ft.-wide traffic lane between it—and all of the perimeter cabinetry—and the nearby island is wide enough for a wheelchair to pass through.

In this universal kitchen, microwaves, ovens, refrigerators, and sinks are all doubled up, making it work well for two cooks at one time. Meanwhile, cabinets have their own ADA touch: Higher- and deeper-than-usual toe kicks allow space for wheelchair footrest.

While a standard-height sink is positioned beneath a window, its ADA-height counterpart is recessed into the island—complete with wheelchair accessibility beneath it. Throughout the kitchen, non-skid vinyl flooring with a wood-grain look allows wheelchairs to easily navigate the space.

A New Approach to Work Zones

The most efficient kitchens were planned, for years, around a single home cook. The range, sink, and refrigerator served as points of an imaginary triangle, with the total distance between the three points adding up to no more than 26 ft. overall, and each leg measuring between 4 ft. and 9 ft. That concept still works well in many kitchens but others have expanded in size, adding more workstations along the way. Maybe, instead of a single range, your kitchen features a separate cooktop and wall ovens. Or a primary sink is supplemented by a prep or bar sink. Even a refrigerator might be split into a separate fridge and freezer, or supplemented by refrigerator drawers. When separate or secondary appliances are put into play, you may find that you have two distinct work triangles but if so, it's still important to keep workstations in each no more than 4 ft. to 9 ft. apart.

PRIMARY WORK ZONES

A kitchen's primary work zones remain unchanged, providing the hardest-working areas:

Cooking: Encompassing more than a stove—or cooktop and oven—cooking zones might also include a vent hood, a microwave, convection, or steam oven, as well as a pot filler.

left · In a one-cook kitchen, the traditional work triangle often works best. The range, sink, and refrigerator should serve as points of an imaginary triangle, with the total distance between the three points adding up to no more than 26 ft. overall, and each leg measuring between 4 ft. and 9 ft.

left • There are very specific zones in this kitchen—the cooking zone, with its range and hood; the refrigeration zone, to the right of the range and defined by contrasting cabinetry; and the clean-up zone under the windows. In addition, the spacious island does double duty as a food-prep and casual dining zone.

Clean-up: The primary elements of the prep/clean-up area are typically the main sink and standard dishwasher. But it can also incorporate a prep sink or dishwasher drawers plus trash and recycling receptacles.

Refrigeration: The refrigerator/freezer is the main player in this zone, although it might include refrigerator or freezer drawers and a beverage cooler, too.

SECONDARY WORK ZONES

Depending on your lifestyle, secondary work zones may be just as important to you as the primary areas.

Baking: An avid baker may want a specialized baking zone, with a convection oven, mixer lift, and a marble countertop for rolling out dough.

Home office: A designated home office tucked into the kitchen might be used for managing bills, the shopping list, and the family calendar. With a work surface large enough to accommodate a computer, it can be the perfect place for kids to do their homework, too.

Kid-friendly: More and more families are creating areas for their children to safely use kitchens. Refrigerator, freezer, and microwave drawers—as well as beverage coolers—keep appliances within easy reach so kids can help themselves.

below • The traditional concept for work triangles works, too, if you have a separate cooktop and ovens instead of a range. Ideally, however, they should be adjacent or at least very close to one another.

Get in the Zone

The primary work zones in this contemporary kitchen are precisely where you might expect them to be. The main sink and refrigerator are on opposite sides of the room, while the range is positioned between them. But secondary work zones are equally important to this family's lifestyle. To the left of the sink, a built-in coffee system is surrounded by plenty of open storage that keeps mugs and espresso cups close at hand, and is easily accessed without getting in the way of the cook. Meanwhile, between the refrigerator and the barn-style pantry door, an appliance garage keeps a blender and toaster oven within easy reach, though they can be hidden at a moment's notice behind a tambour door. Finally, just a few feet beyond the kitchen proper, a wine cooler is convenient when entertaining is on the agenda. Work zones devoted to specialty appliances like these can take an ordinary kitchen into the realm of extraordinary.

above · The coffee system in this kitchen is well thought out, with plenty of open shelves for dinnerware—including coffee cups—as well as favorite cookbooks. Behind doors and in drawers, there's more space to stash all manner of coffee accoutrements, among other things.

left · Storage space in the dining area accommodates more than serveware and linens; there's an easily accessible wine cooler, too, keeping bottles at just the right temperature. The countertop above is the ideal spot to pour wine—or set up as a buffet.

above • It's no coincidence that two of the work zones in this kitchen are located at the edge of the space. The coffee system is on one side, making it easy to grab a cup in the morning without getting in the way of the cook. The same concept applies to the refrigerator; snacks and beverages are easily accessible anytime.

left • In the open floor plan of this home, the formal dining area is adjacent to the kitchen. The same contemporary cabinetry is used in both spaces, tying them together seamlessly.

The Importance of Islands

If increased functionality is the goal in your kitchen, an island offers tremendous potential. It can conceivably be an all-in-one work zone, with fixtures, appliances, and storage all in a single step-saving spot. Or, it might serve as a specialized work center; a marble-topped island, for instance, would be perfectly suited for the baking enthusiast.

When adding an island to an existing or brand-new kitchen, allow enough floor space for the traffic paths around it; 42 in. is recommended for a single-cook kitchen, while 48 in. is best for two cooks. Don't think that your island has to match the rest of the kitchen, either. Complement the rest of the room's cabinetry by using a different wood, or the same wood in a different color or finish. Or, change things up by using a different material entirely than what's on the perimeter cabinets.

If you're giving your current kitchen a facelift, you might opt to simply upgrade your existing island. An entirely new countertop, for instance, can give your kitchen a fresh new look. You can build in a spot for casual dining, too, by extending the countertop 15 in. along one side. Just be sure that the overhang has sufficient support.

While working at this expansive island, the cook has a clear view outside the bank of windows. The edge closest to the windows serves its own purpose, creating one "wall" of a walkway between rooms.

above • Many an island accommodates closed storage, casual dining, and even a sink. But this one goes a step further to include open shelves that hold a collection of cookbooks, accessible in an instant.

right • Substantial islands seem to be the norm more often than not, but this kitchen proves that's not always necessary. This scaled-down version still meets every need—a place for food prep, a landing spot for the range, and a casual eating area.

Because it's so important that stools are a good fit for the height of an island, the ones in this kitchen are the ideal choice. Their adjustable height allows the occupant to raise or lower the seat to the perfect position.

Eat-In Kitchens

The possibilities for in-kitchen dining areas run the gamut from a small breakfast bar to a standard table. But no matter what option you choose, give careful thought to its placement. If your eating area is too close to the food prep zone, for instance, diners may get in the way.

BREAKFAST BARS

The breakfast bar is, in fact, a great place to dine any time of day. Side-by-side seating is ideal for diners who want to see what's cooking and to chat with the cook. Positioning the stools in an L-shape, though, results in more camaraderie; diners can converse with ease and still watch what's going on in the kitchen. But whatever configuration you choose, a breakfast bar should be at least 18 in. deep and between 36 in. and 42 in. high. And don't just buy stools on the basis of looks; it's critical to match the height of your counter to the stools and vice versa (see "The Importance of Elbow Room" on p. 51 for specifics). Keep in mind, too, that stools are great for sliding under a countertop when not in use, but those with backs will give you more support.

The lattice-back stools in this kitchen are not only good looking but also built for comfort. The contemporary pieces have slightly curved, supportive backs as well as matching seat cushions.

The countertop in this modest kitchen turns a corner and extends just far enough to create a simple peninsula for casual dining. If extending the countertop isn't an option, you could get the same effect by placing a table at a right angle to the kitchen counter—as long as the two are the same height.

A pair of wooden stools in this country kitchen slide neatly under the countertop when not in use. And because they're lightweight, they can be moved elsewhere in the room when—and if—they're needed.

BUILT-IN BANQUETTES

Built-in dining can be designed like a booth, with two benches facing each other, or with benches in an L- or U-shape. For booth and U-shape seating, it's important to choose a pedestal table that won't interfere with diners getting in and out, while a four-legged table can be used with L-shape seating. Either way, be sure that the table you choose is lightweight, so it can be moved easily for cleaning. Keep in mind, too, that a built-in banquette will most likely be used for more than eating. If, for instance, it also serves as a homework spot, be sure to provide electrical outlets and perhaps a charging station.

above · When it comes to in-kitchen eating areas, flexibility can be your biggest asset. This freestanding table, for instance, can be moved to wherever it's needed most. As for the seating, the upholstered wingbacks can be teamed with more or fewer metal stools as needed.

left · The beauty of a banquette is that it can be tucked neatly into a corner and the table can be pulled out for easy cleaning. Additionally, extra chairs can be pulled up to the table as needed.

FREESTANDING TABLES

When it comes to in-kitchen eating, a simple table and chairs combination is still—for many—the go-to approach. Their flexibility holds much of the appeal; chairs can be pulled away or added as needed. If your kitchen has enough square footage, the best solution may be a combination of built-in and freestanding dining areas that can adapt to your changing needs.

Many an in-kitchen eating area isn't in the kitchen at all; rather, it's just a few steps away, so dishes can be served and cleared easily. The splayed-leg design of this table dictates that chairs be pulled up to the sides. And because the chairs are side by side, an armless style is the best option.

The Importance of Elbow Room

To allow ample space for a seated diner, the National Kitchen & Bath Association recommends 32 in. between the table or countertop edge and the nearest vertical wall or obstruction—if no through traffic passes behind him or her. If, however, traffic does pass behind the seated diner, that dimension should be increased to 44 in. For countertop height, depth, and width, see the chart below.

	COUNTER HEIGHT	KNEE-SPACE DEPTH	WIDTH PER SEAT	SEAT HEIGHT
TABLE DINING	28 in. to 30 in.	18 in.	24 in.	18 in. to 19 in.
STANDARD COUNTERTOP HEIGHT	36 in.	15 in.	24 in.	24 in. to 26 in.
BAR HEIGHT	42 in.	12 in.	24 in.	30 in.
UNIVERSAL DESIGN ACCESS FOR WHEELCHAIR	27 in. to 34 in.	17 in. at feet; 11 in. at knees	36 in.	N/A

Have a Seat

Whether it's for mealtime, homework, or casual entertaining,
family and friends are congregating in the kitchen more than ever
before. Not surprisingly, then, comfortable seating is key.

above • This dining area, adjacent to the kitchen, cleverly uses the island to create a back for the bench. There's drawer storage beneath the upholstered seat, but it's easy to access because the table and chairs can easily be pushed away.

facing page • Although counter stools are the norm for dining at an island, a pair of upholstered benches is a good fit for the traditional room—as long as they're the right height. And because this kitchen is part of an open floor plan, they can be moved wherever they're needed.

above and left • A kitchen with diminutive dimensions usually can't accommodate seating, but this one does the next best thing by utilizing a short wall just around the corner. The taller-than-standard table requires higher seats, too, provided by a pair of stools and a built-in bench that's been raised a few inches and given a footrest for an extra measure of comfort.

CABINETRY

Cabinetry gets top billing in a kitchen, not only because it sets the style and shapes your space; but also because it demands a fair share of your budget. So it's important to consider all of your options carefully, right down to the last knob or pull.

Know the Basics

When shopping for cabinetry, start by getting comfortable with cabinet lingo, starting with cabinet case construction; do you prefer a face-frame or a frameless case? Educate yourself on door and drawer types, as well as hardware and cabinet accessories. And get creative with configuration. Not all base and wall cabinets have to match, and you may not want to fill an outside wall with cabinets if they will block a great view. Once you've thought through every last decision, you're ready to shop.

It's important to keep a tape measure close at hand—and have a clear vision of your new kitchen. Standard wall cabinets are 30 in. high, and, if set at the suggested 15 in. to 18 in. above a 36-in.-high countertop, they won't reach the ceiling. The space above the wall cabinets might be used for display, or the ceiling can be dropped to make a soffit. But it's currently more popular to take cabinets all the way to the ceiling, either by installing wall cabinets higher, ordering taller cabinets, or using a combination of standard cabinets topped by transom cabinets.

Base cabinets are set on solid bases or on legs, which can either be exposed or covered by trim; many feature recessed toe kicks, too. Stock face-frame and numerous frameless manufactured cabinets have a 4-in.-high, 3-in.-deep toe space, while European-style cabinets more often have tall toe spaces, measuring from 5 in. to 8 in. In addition to making it more comfortable to work close to the countertop, a toe space can serve other purposes. It might be fitted with a pull-out drawer that holds a step stool or provide an inconspicuous place for an air register.

left · The traditional approach for wall cabinets is to place them side by side, but this contemporary kitchen handles them in a contemporary way. They're spaced intermittently on two walls, allowing breathing room around the kitchen's window and cooktop's hood.

below · This kitchen relies on base cabinets to provide all the closed storage it needs, not only beneath the bank of windows but also on the adjacent wall. The latter incorporates shelves, as well, making it easy to grab oft-used pieces quickly.

left · If set at the suggested 15 in. to 18 in. above a 36-in.-high countertop, wall cabinets won't reach the ceiling. While the tops of cabinets can be used for display, it's more popular today to run cabinetry all the way to the ceiling. This kitchen, for instance, combines standard and transom cabinets.

Construction Matters

How cabinets are assembled, and from what materials, has a major impact on how well they will stand up to the wear and tear of a busy household. Cabinets come in three basic materials, typically not made entirely of solid wood but, instead, of wood-based materials. That's because solid wood can absorb moisture easily and is prone to warping and cracking if not properly sealed. Wood-based materials, on the other hand—such as plywood, medium-density fiberboard (MDF), and particleboard—are not. Thus, cabinets advertised as being constructed of "solid wood" are usually made of engineered wood materials covered with veneers such as maple, oak, or cherry.

For years, the highest quality cabinet cases have been made from ¾-in. veneer-core plywood, which is stronger, lighter weight, and more moisture resistant than MDF or particleboard. On the other hand, some cabinetmakers prefer MDF over plywood for its dimensional stability and its smooth face that's ideal for applying veneers and other laminates (it's also typically less expensive than veneer-core plywood). And while it's true that particleboard is the lowest quality of case goods, it's also the least expensive—and the most commonly used in stock cabinets. If your budget allows, choose plywood for cabinets where water could potentially cause damage. For dry locations, look to MDF or even particleboard, but also consider a plywood mashup, combi-core; it has a strong and light veneer plywood core sandwiched between layers of MDF, providing a smooth, stable surface.

Dovetail joints—one of the strongest types—command a higher price than their stapled counterparts, but the quality they provide is well worth it.

Shelves

With endless possibilities, open shelves continue to gain popularity in the kitchen. More and more often, for instance, they're taking the place of wall cabinets, not only making items more accessible but also giving the room a more spacious feeling.

Selecting the best shelf material comes down, in large part, to aesthetics. Solid wood is relatively strong, but it can warp. Plus, it expands and contracts with changes in the humidity. Veneered plywood, on the other hand, is not only more stable than solid wood but

As a rule of thumb, a shelf should be just slightly deeper than the objects displayed on it.

also can pass for solid wood if its edges are covered with a glue-on or iron-on edging or edge-band (which also helps stabilize the shelf). Although MDF and particleboard can't span as far as solid wood or plywood shelves of the same thickness and depth, they're certainly serviceable if their supports are close enough together to prevent sagging.

Custom cabinetry can do more than deliver the quality of case that you prefer; it also allows you to choose your preferred configuration, color, and finish, exemplified by these blue lacquered examples.

Where cabinetry materials will be used will greatly affect your choice, and how you will use them matters, too. In order to support the heavy marble top of this island, for example, the legs need to be of the highest quality.

If the overhang of an island isn't too heavy, it can be supported by a simple bracket instead of a full-fledged leg. This approach has the advantage of interfering less with the counter stools.

Wood Cabinets

Wood cabinets are preferred by many homeowners, but there are numerous varieties from which to choose. Because wood is a natural product, it will vary greatly in color, texture, and grain. Plus—with exposure to sunlight—the appearance of some woods can change over time.

CHERRY

With a medium reddish-brown color and a uniform grain, cherry is one of the most popular woods used in cabinetry. Although more expensive than some other types, this hardwood (though softer than oak or maple) resists warping and cracking, too, making it a good choice for kitchens.

MAPLE

Maple ranges from creamy white to pale reddish brown in color, with its smooth, uniform appearance making it ideal for paint or stain. In fact, most painted cabinets use maple as their base.

BIRCH

Like maple, birch has a straight, even grain and a golden appearance. It is sometimes used as a more affordable substitute for maple.

OAK

A durable hardwood well suited to traditional, casual, or rustic looks, oak is light brown in color with a strong, open grain. It's typically characterized by a cathedral-like pattern but can be rift-cut or quartersawn for a more even appearance. Be aware, however, that these options can cost considerably more than standard oak and are rarely available in stock cabinets.

Stained cabinets in this kitchen are a good match for the room's natural pine walls and ceiling. The granite countertops and backsplash complement the cabinetry while inspiring the dark knobs and pulls.

Cabinets and Finishes

Oak, cherry, pine, hickory, and maple are some of the best choices for traditional cabinetry. But they can adapt to contemporary styling, too, while mahogany and bamboo are often reserved for contemporary cabinets. Most wood cabinets are sprayed with catalyzed varnish, although hand-applied finishes provide a true vintage look. Be aware, however, that hand-applied glazes and paints, as well as high-gloss lacquer, are the most costly finishes. And painted cabinets—especially those with glossy finishes—show dings more easily than clear-finished wood cabinets, making them more difficult to touch up.

above · Straight-grain woods or veneers, especially those that are light in color, are a good fit for a contemporary kitchen. Here, they contrast beautifully with the dark hood and island countertop, creating a dramatic atmosphere.

left · With beaded façades and a dark stain, these doors and drawers are undeniably traditional in style. Drawer pulls on all give them a cohesive look, and they can be easier to use than knobs.

PINE

Pine, a budget-friendly wood material, can provide a distinctively rustic look—popular in many a traditional or country kitchen. Pine is a softer wood, however, and may dent or scratch more easily than some other types.

MAHOGANY

Known as the premier wood for fine cabinetry, mahogany has a rich, reddish-brown color and a characteristic swirling grain. Because this species of wood has become increasingly rare—and because it's extremely durable—mahogany is significantly pricier than its more common counterparts, such as pine and oak.

White-painted cabinetry has been popular for years, but today the color palette is expanding. Grays, greens, and blues are more and more prevalent, often applied to maple as that's the wood that takes paint or stain most easily.

ALDER

Sometimes used as a more affordable alternative to cherry in semi-custom and custom cabinets, alder characteristically has reddish-gold tones though alder facades are less dramatic in contrast than cherry. Alder is softer than either maple or cherry, but it's still a stable surface for stains and finishes.

HICKORY

With extreme color contrasts and an open grain, hickory is extremely hard. In terms of appearance, however, it's prone to mineral streaks, bird pecks, and other evidence of its past life as a tree.

above • Wood cabinetry has a warmth all its own, but the reddish-brown tone of cherry makes it even more so. The straightforward design of these cabinets allows the wood itself to take the starring role.

right • Part of the beauty of maple cabinets is that they're right at home in any kitchen. Depending on the door and drawer styles, the countertops, and the hardware, they can take on traditional, transitional, or contemporary characteristics.

Bamboo Cabinetry

Although often mistaken for wood, bamboo is a grass that's both strong and sustainable. And because it grows to maturity in as little as seven years—making it available more rapidly than wood—it's become increasingly popular for cabinetry.

Bamboo's sturdiness makes it particularly well suited for high-traffic areas like the kitchen, and its natural color is so appealing that most homeowners don't feel the need to stain it; bamboo needs nothing more than a simple clear coat. Plus, the grain of bamboo is distinctively linear and can be oriented either vertically or horizontally. Because bamboo is a fairly new and limited resource, it's still somewhat expensive; bamboo can be comparable to the cost of high-end hardwood. Stock styles are particularly limited; you may need to order custom cabinets.

Laminate, Lacquered, and Metal Cabinets

There's no need to think of plastic laminate as a second-rate substitute for wood. Today's options are just as stylish as they are sturdy. Not only are look-alikes of common woods available; so are book-matched finishes inspired by exotic woods that would otherwise be beyond your budget or simply unavailable. There's also a price incentive: Laminate cabinets cost approximately 15% to 30% less than their wood counterparts. Laminate doesn't come without its drawbacks, however. Because it isn't malleable, it's most effective on slab cabinet fronts; the material doesn't lend itself to ornamental doors or applied moldings. If damaged, laminate cabinets can't be refinished, either, or even easily repaired.

Lacquered cabinets are characterized by their reflective finish, one that's particularly well suited to contemporary kitchens. Flat-panel cabinets are the best application, though the lacquer itself—available in a wide range of colors—can be either matte or high-gloss. Because the finish is factory-applied, lacquered cabinets are durable but they're expensive, too. They do, however, typically come with an excellent warranty.

All the rage in the 1930s and 1940s, metal cabinets are also enjoying a resurgence today; a new generation of homeowners is discovering how affordable and low maintenance these cabinets can be. Resistant to both stains and corrosion, they're easy to clean. Plus, metal cabinets are nontoxic, making them ideal for people with chemical sensitivities. Don't think that conventional stainless steel is your only option, either; shiny as well as brushed metals are available in stock, semi-custom, and custom lines. You can even find powder-coat stainless-steel cabinets in a variety of colors.

above and right • Some of today's finest laminates are inspired by—and a near replica of—exotic woods. Best of all, they typically cost a fraction of the price. A wood-lookalike laminate, for instance, wraps the island of this contemporary kitchen and covers the base cabinets. But the room does an about-face on the other side, where the storage wall has a shiny white laminate finish. The combination of the two is smart: It doesn't allow the room to get weighed down in a dark wood tone nor does it become sterile in all white.

left • Metal cabinets in this kitchen give it a clean, appealingly retro look—a welcome update from their introduction decades ago. Beyond their appearance, there's another plus: Because they're nontoxic, they fall into the category of being "green."

Mixing Things Up

Just as matched suites of furniture are no longer a necessity in the rest of the house, nor are identical cabinets in the kitchen. In fact, a kitchen that mixes colors and textures can be more visually interesting—and offer a clue to your personality, too. You might, for instance, turn your island into an accent piece; it's one of the easiest ways to incorporate a second color into your kitchen design. You might spice up a clean, contemporary space with a pop of color or complement wallpaper or paint in a traditional room. On the other hand, the island isn't your only option. Wooden hoods, pantry doors—even statement appliances—can insert colorful accent colors.

As an alternative, you might opt to use different colors on your wall and base cabinets. To ground the overall design, choose a darker shade for your base cabinets, then turn to lighter hues—like whites and grays—to keep the color scheme from seeming top-heavy. The result will not only be more visually interesting but also makes your kitchen feel more spacious.

Finally, don't hesitate to play with texture. Wood and metal, for instance—perhaps in the form of butcher-block countertops and stainless-steel appliances—can team up with any combination of colors to create a one-of-a-kind kitchen design.

above • Sometimes all it takes is a touch of color to make a big difference in a room. This collection of green bowls and bottles wouldn't have been nearly so impressive against a white backdrop. Papering the back wall in a red-and-white print makes all the difference, taking the look from ordinary to extraordinary.

right • A commercial-style range can accommodate every imaginable cooking task. But a stainless-steel finish is no longer your only choice; they're now offered in a wide range of colors, ready to add some individuality to your kitchen.

Although colors and textures mix it up throughout this kitchen, it's the canary yellow island that's the focal-point piece. In the midst of white cabinets throughout the rest of the room, it adds a cheerful touch, while the navy counter stools provide yet another layer of color.

Taking a cue from the adjacent kitchen, the dining area mixes things up, too. Banquette seating accommodates a Kelly green rattan table on one side, while—on the other three—blue metal chairs are almost an exact match for the range.

Face-Frame vs. Frameless Cabinets

Cabinets start with the basic box—often referred to as the case or carcase—which can be either face-frame or frameless. Both have their pros and cons; it's a matter of what suits your style and your budget.

FACE-FRAME CABINETS

A face-frame cabinet case gets its style and strength from a frame of horizontal rails and vertical stiles applied to the exposed edges of the case. Doors mount to that frame, either fitting flush into the frame or overlaying all or part of it. Because it takes more time to construct components that must fit closely together, face-frame cabinets with inset doors and drawers are pricier than those with overlays. Partial-overlay face-frame cabinets are less expensive than full-overlay doors, simply because there's a wider gap between the doors and drawers.

FRAMELESS CABINETS

A frameless cabinet case—also known as a European cabinet—is a box with no face frame. Because there's no frame to add stability, the case itself must be built stronger than a face-frame counterpart; ¾-in.-thick sides make the sturdiest frameless case. From the outside, it's not always easy to distinguish face-frame cabinets from frameless; doors and drawers for both can be flush overlay.

You'll find a vast assortment of stock and semi-custom cabinets, or you might opt to specify your own custom cabinetry. If you're a do-it-yourselfer, and trying to keep costs down, consider knockdown (KD) and ready-to-assemble (RTA) options, too.

above • Frame-and-panel doors immediately establish a conventional feeling in this kitchen. But the blue cabinets, coupled with a crisp white tile, offer a fresh twist on tradition. Inset doors and drawers like these are typically available in custom cabinetry.

right • Flush overlay doors and drawers are a good choice for a contemporary kitchen; their clean lines are in keeping with the style's sleek, streamlined approach. In lieu of standard hardware, integrated handles can take the look to the next level.

Face Frame/Frameless

THE CABINET CASE

- A face-frame cabinet is easier to fit into a space that isn't completely square and plumb.
- A face-frame cabinet has a narrower opening than a frameless cabinet of the same width, so pull-out shelves and drawers will be narrower, too.
- A frameless cabinet has no stile or rail in front of the contents, making it easier to pull out stored items; an exception is an especially wide cabinet that may require a center post.
- A face-frame cabinet gets much of its strength from the frame, whereas a frameless cabinet depends on a stronger, thicker back and strong corner joints.

DOORS AND DRAWERS

- In frameless cabinets, doors and drawers usually overlay the case completely—referred to as full overlay or flush overlay. Frameless cabinets rarely have inset doors.
- In face-frame cabinets, doors and drawers may overlay the frame completely, may be inset, or may overlay the frame partially (referred to as reveal overlay or half overlay).
- Inset doors, the standard in traditional-style cabinets, typically require more precision in their construction and installation than overlay doors.

DOOR HARDWARE

- Concealed adjustable hinges are available for both frameless and face-frame cabinet doors. They commonly adjust in three directions and are easy to tweak over the lifetime of a cabinet.
- Inset doors, used almost exclusively in face-frame cabinets, are typically hung with butt hinges, which require more precision to install than their adjustable counterparts.

Face-frame cabinet

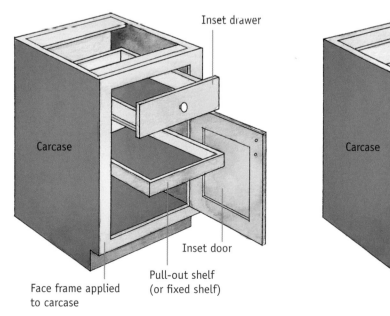

Inset drawer

Carcase

Face frame applied to carcase

Pull-out shelf (or fixed shelf)

Inset door

Frameless cabinet

Full overlay drawer

Carcase

Full overlay door with adjustable hinges

What's Your Type?

Your cabinets' components are likely to come from a variety of places. Even custom cabinetmakers may source out components to specialized suppliers and then assemble the cases in their own shops. Stock and semi-custom cabinets increasingly come from one of the many cabinet manufacturers that purchase cabinet parts from companies that specialize in doors, drawers, or cases. And you can even mix and match yourself. You might opt for custom or semi-custom cabinets for high-visibility locations, such as the island, and use stock cabinets for perimeter locations.

STOCK

$

- Can be purchased right off the shelf or ordered from a big-box store, home center, or lumberyard, or through a kitchen products dealer, designer, or contractor; installation is often available for an additional fee.
- Standard-size components are typically built in 3-in. increments; if a run of cabinets isn't quite as wide as you need, stock fillers can span the gaps.
- Limited finishes include oak, maple, and cherry veneers, as well as some basic laminates.
- Typically, the most affordable option, running about half the cost of semi-custom and custom cabinets.

SEMI-CUSTOM

$$

- Available in a wider range of styles and finishes, including oak and quartersawn oak, cherry, maple, birch, hickory, and painted finishes.
- Wider selection of accessories and moldings.
- Cabinets can be built as larger assemblies instead of case by case.
- Semi-custom cabinetry tends to be higher quality and higher priced than stock cabinets, sometimes by a considerable amount.

CUSTOM

$$$

- Widest selection of style and finish offerings, including exotic woods and high-end laminates.

Depending on the intricacy of your plan, white perimeter cabinets can be executed with stock, semi-custom, or custom units. This reclaimed wood island, however, is a custom piece, incorporating just enough shelf space at each end for cookbooks while leaving room for three counter stools.

- Cabinetmaker may combine components from several specialized sources with shop-built cases, which can translate to shorter lead times and better products.
- Can be built to specific measurements and dimensions.
- Typically the most expensive cabinet option.

Between this refrigerator and freezer, frame-and-panel doors and drawers appear to be standard cabinetry. But behind a set of tall doors there's a breakfast bar, where coffee can be brewed, smoothies made, and breakfast items warmed up. Because this configuration—including the cabinets throughout the kitchen—requires specific dimensions, a custom approach is best.

Flat slab doors and drawers like those in this contemporary kitchen can often be executed with stock cabinetry, typically the most affordable option you'll find. But you can give it a custom look with the wide array of hardware that's on the market today.

Drawers and Doors

Doors and drawers are made in two basic types: Frame-and-panel doors and drawer faces are more traditional in style, with a frame often made of solid wood and panels that are solid or veneered MDF. Flat-slab (one-piece) doors and drawers, on the other hand, are made from glued solid wood or MDF veneered with any number of materials, most commonly wood but also plastic laminate, metal, or even glass.

When selecting doors and drawers, keep in mind that inset styles—which fit flush with face-frame cases—cost more than overlay designs due to the extra precision that's required. You'll find that hinges on inset cabinets are visible, with mortised butt hinges and leaf hinges being the most traditional styles.

Overlay doors and drawers affix to the surface of a face frame or the interior of a frameless case, with cup hinges making them easily adjustable. Full overlay doors and drawers, the standard on frameless cabinets but also used on face-frame cabinets, all but touch each other. Thus, they're more painstaking to build and install than reveal overlay doors and drawers, which are spaced farther apart. Reveal overlay doors and drawers—also called partial overlay—are used on face-frame cabinets.

above • Partial overlay doors and drawers front the cabinets in this kitchen, but—with different colors and hardware—the wall and base units look nothing alike. The white wall units, in a mix of solid and glass-front doors, feature simple brass knobs, while navy blue base cabinets primarily feature bin pulls, with a brass knob here and there.

right • Inset frame-and-panel drawers at the end of this island may be traditional in style, but their color makes them anything but. They have a little extra detailing, too, their beaded edges providing a framed effect.

A higher-than-usual ceiling—complete with a skylight—allows the center section of these cabinets to step up, flanking the hood of the range and topping it, too, for that matter. The inset doors have a clean look, allowing the black decorative hardware to be showcased.

Glides

Drawers are operated by glides, which can be side-, bottom-, or corner-mounted. Full-extension glides allow access to the entire length of a drawer, an extra expense that's well worth it, for example, in large drawers that hold pots and pans. Quiet, self-closing glides are another feature you may want to spring for; they retract with a gentle nudge, preventing drawers from slamming shut. Some glides pause at various marked positions, while pull-through glides can be opened from either side of an island.

Glides of all types are made from plastic or metal ball bearings and vary in the amount of weight they can bear. Three-quarter-extension, epoxy-coated glides, for instance, can support up to 75 lb., whereas full-extension glides, which allow access to an entire drawer, can hold up to 100 lb. Before committing to your drawers, be sure to check their load ratings.

DRAWERS

It goes without saying that drawers do some heavy lifting, but that applies today more than ever before. Base cabinets are now often fitted entirely with drawers instead of drawer-and-door combinations. This approach makes storage more accessible, allowing you to see—and reach—the contents more easily.

Be sure, however, that your drawers are up to the task. Because they'll be supporting stacks of dishes, as well as pots and pans, look for drawer boxes that are sturdy; they should be built ½ in. to ¾ in. thick, especially on the bottom. Solid hardwood has long been the material of choice for drawer box fronts and sides, fastened with dovetail, finger, tongue-and-groove, dowel, biscuit, or dado joints. But drawer boxes are also commonly made from engineered woods—such as veneered plywood—metal, melamine, particleboard, and MDF. Plywood, on the other hand, can be a better option for drawer bottoms because it won't shrink or swell.

above • Flat-panel doors and drawers are a good fit for a contemporary kitchen, and this black-and-white color scheme maintains the modern theme. Keep in mind, though, that simple styling doesn't necessarily mean the cabinets are less expensive; frameless cabinets like these can be more painstaking to build and install.

left • Frame-and-panel doors are right at home with flat-panel drawers in this kitchen, their gray-green color and black hardware tying them together. Adding to the consistency is the fact that they are all inset.

Hinges

When it comes to hinges, there are more decisions to make than you might think, depending first on whether your cabinets are face frame or frameless. In face-frame cabinets, hinges typically mount to the face frame while, in frameless cabinets, they mount to the interior. Also to be considered is the degree of opening that you need. Some hinges allow a door to open just beyond 90°, whereas a hinge that permits a 270° opening lets a door swing back against the cabinet side (though this type of hinge only works on frameless cabinets with full overlay doors).

Finally, different hinges can have a huge effect on the appearance of your cabinets and doors. Concealed hinges—not visible when cabinet doors are closed—are a good choice for contemporary kitchens, where hinges on sleek cabinetry would be a detraction. Semi-concealed hinges are partially visible while exposed hinges are fully revealed; both may have decorative detailing. Butt hinges, on the other hand—either non-mortised (surface mounted) or mortised (set into the case)—are traditionally used for inset doors. Installing mortised butt hinges requires extra time and precision but creates a smooth fit between the door and the frame.

DOORS

Your cabinet's door style makes a huge impact on the overall look of your room. While modern, Euro-style cabinets will lend a contemporary feeling, ornate doors and drawers—especially with a distressed finish—will create an unquestionably traditional look. If you're opting for custom cabinets, the sky's the limit in terms of design styles. Even semi-custom cabinets offer plenty of options, but with stock cabinets your choices are limited.

That said, cabinet doors—to a large degree—have common characteristics. Raised panels refer to center panels that protrude from the door frame, while recessed panel doors feature center panels set inside the frame. And it makes sense that flat panel doors have no profile at all; their facades are totally smooth.

A tall cabinet in this kitchen, with its recessed panel doors and drawers, carries out the cohesive design of the room. Behind the façade, though, it's hiding a secret: The unit opens to reveal a spacious pantry, fitted with the same cabinetry as the kitchen.

Reface or Refinish?

New cabinets can eat up to 50% of a kitchen remodeling project. But perhaps you're trying to keep a lid on your budget. Or maybe you just want to go "green" by keeping your current cabinet boxes out of the landfill. If they're in sound structural condition, refacing or refinishing the facades may be the best solution. But what, you may ask, is the difference?

• *Refacing* involves adding a layer of laminate or wood veneer over existing cabinet surfaces; by changing the style, the wood species, and even the color, you can give your cabinets a complete facelift. Refacing can save up to 50% compared to the cost of new cabinets. And because it doesn't require the removal of appliances, the kitchen can still be functional while work is being done.

• *Refinishing*, whether it's painting or staining, can be an even less expensive way to bring cabinets back to life. The process requires that you first clean the doors, shelves, and hardware, then sand and/or strip any old paint, blemishes, and flaws from the surfaces. Once that's done, you're ready to paint or stain your cabinets for a fresh, new look.

Whether you opt to reface or refinish, new hardware can further enhance your "new" cabinets. Exposed hinges can be re-plated or replaced, while knobs and pulls can also be updated. Likewise, new interior accessories can greatly enhance the functionality of your cabinetry.

Painting cabinetry is one of the best ways to refinish; it can bring tired cabinets back to life and add personality, too. Shades of blue and green now complement the original wood cabinets and trim in this kitchen, adding a spirited aspect to the space.

above · Refacing with wood veneers gives you the capability to change the species of your kitchen cabinets, even to something exotic that would otherwise be out of your budget's reach. Flat-panel cabinets like these are some of the easiest to reface.

left · Whether you opt to reface or refinish, replacing hardware can take your transformed cabinets to yet another level. When changing out pulls, however, be sure that the holes drilled for the old hardware match up with the measurements of the new.

Interior Accessories

Cabinet accessories can be built in during the manufacturing process or purchased as part of a stock or semi-stock cabinet package. In addition, you'll find all kinds of after-market accessories online and in big-box stores.

Well-organized cabinetry can make a huge difference, allowing you to sort, store, and access your kitchen items in fewer cubic feet. Some accessories simply divide space; consider built-in vertical slots for flat items such as trays and cutting boards. Other accessories, meanwhile, are mechanical, sliding out, up, or down. If you prefer to keep your stand mixer out of sight until baking time, an appliance garage is one option; those with lift-up, tilt-in, or swing doors are particularly good choices. But a spring-up shelf also allows you to store a mixer or other heavy appliance under the counter until you need it, while pull-out shelves can handle small appliances, too. Blind cabinet corners can be a challenge, but that doesn't mean there aren't multiple solutions. Look for a lazy Susan, for instance; one without a center post offers more flexible storage, whereas one with wire shelves has better visibility.

Cabinets outfitted for trash, compost, and recycling can be a real asset, too. Contrary to what many people think, the cabinet under the sink is not the best place to put trash and recycling. Instead, locate pull-out garbage bins near food prep areas, to the right or left of the wash-up sink—or both of those work zones. (A pedal-operated trash cabinet allows the cook to dump trimmings without ever touching the handle with dirty hands; another option is a trash bin attached to a door panel operated by a touch latch or, more simply, hooked open with your foot.) Compost bins can be mounted under a countertop hole, as a pull-out or even a built-in drawer. Keep in mind, however, that compost bins must be easy to remove, clean, and replace daily.

What appears to be a door on the exterior pulls out to divulge three pull-out shelves, each one slightly deeper than the next. Attached to the interior of the door, there's even room for a roll of paper towels.

Composting is, more and more, a common practice, and this kitchen has a drawer built for that specific purpose. The top lifts off, making it easy to access and dispose of its contents.

above • Well-placed dowels in this drawer allow efficient storage of dinnerware, while keeping them from rattling around, too. If you're planning to store stacks of plates and bowls in a single drawer, be sure it's sturdy enough to support the weight.

right • Taking advantage of what could have been a hard-to-access corner, shelves swing out from behind a standard cabinet door. Short side rails ensure that items don't fall out when moved back and forth.

right • A sliver of space next to the oven is just big enough for a slim pull-out cabinet. Fitted with pegboard and adjustable hooks, it provides convenient storage for pots and pans of all sizes.

below • Kitchen utensils don't need to clutter countertops or drawers. This pull-out cabinet—directly below the flatware drawer—accommodates canisters that keep spoons, whisks, and spatulas close at hand.

above · Pull-out trash and recycling bins, built into cabinets next to the sink, are convenient and—at the same time—eliminate under-the-sink clutter.

left · Large stand mixers can be stored in base cabinets, set on lifts that bring them to counter height when needed to save potential back strain. As an added benefit, this tactic frees up valuable counter space.

Knobs and Pulls

It may seem like a small part of the decision-making process, but the right hardware can make a big impact on your kitchen's sense of style. Think of pairing cabinets with knobs and pulls like fashion accessories. You can perfectly match your pieces or go for an eye-catching contrast. Similarly, with hardware, one approach is to choose a single finish; then again, mixing finishes can be an indication of your one-of-a-kind style.

When choosing knobs and pulls, it's important to keep the size of one in proportion to the other. A 1¼-in. knob, for instance, is a good size for a standard-size drawer. A knob with a rose—the round plate at the base of the shaft—can keep things neater, simply because fingers are less likely to touch the drawer itself. By the same token, bin pulls have a cleaner look because they're pulled from the inside.

Drawers that are more than 24 in. wide typically require one long pull, or two pulls or knobs; if you opt for two, always use both to keep the drawer from eventually racking. Keep in mind, too, that you can forgo knobs and pulls completely and opt instead for recessed pulls or invisible catches that open when you push on the doors or drawers.

above • The placement of this hardware is such that it doesn't interrupt with the lines of the contemporary cabinetry. With this type of hardware, the extended portion needs to be deep enough to make it easy to grab.

Part of the beauty of pulls like these is that they can be used vertically on doors and horizontally on drawers, bringing continuity to a room.

Nickel hardware beautifully complements the cool blue cabinetry. To handle the width of the drawers, two bin pulls are needed on each, but the adjacent door is slim enough to require only a knob.

Because these brass handles are easy to grasp, they fall into the category of universal design—a plus for any age or ability.

Knobs and pulls not only offer an easy way to embellish cabinets, they're the least expensive, too. These ceramic knobs with their crackle finish add just the right decorative touch.

On this face-frame cabinet, a simple brass knob is all that's needed. Plus, it doesn't visually conflict with the dinnerware behind the glass-fronted door.

Decorative Details Make the Difference

Decorative detailing can take your cabinets from ordinary to extraordinary, whether it's a single element repeated several times or an assortment scattered throughout the kitchen. Here are some of the options to consider:

- **Moldings:** Crown moldings atop wall cabinets can add a strong sense of style to your kitchen. If the moldings are to be used on top of existing cabinetry that you don't plan to refinish, though, it's easier to work with a complementary paint or stain rather than try to achieve an exact match on the wood.

- **Feet and legs:** Because they nearly always match the cabinets to which they are attached, these decorative elements are easiest to incorporate into new cabinetry. If, however, you're replacing a countertop on an existing island—and there's an overhang that needs support—decorative feet and legs can perform that function, while adding a touch of fashion, too.

- **Valances:** With a variety of applications throughout the kitchen, valances come in different shapes and styles. A valance above a cooktop can disguise an exhaust fan, for instance, while a valance over the main sink might hide recessed light fixtures.

- **Corbels:** Although their main purpose is to support heavy countertop overhangs, corbels have a fashion element, too; they can make even the simplest cabinet look stunning. They should, however, be sized appropriately for the dimensions of the overhang and be professionally installed.

- **End panels:** When a run of cabinets doesn't end at a wall, end panels are recommended. By matching the cabinets' doors and drawers, end panels can lend a more finished look.

above · In lieu of a straightforward design, supportive legs on this island take a decorative tack. The pale blue carved legs are prominent, too, against the contrasting dark wood floor.

left · Valances can also be added over a sink, where they can hide recessed light fixtures or a window treatment's hardware.

below left · Crown moldings can dress up traditional and transitional cabinetry. Be sure, though, to choose a style that's in keeping with both the cabinetry and the room.

left · Valances, like the one over this range, disguise the vent while adding a decorative element, too. Components like this are easiest to incorporate if you're doing a complete remodel, so they match the cabinets to which they're attached.

The Perfect Pantry

A pantry can take on many forms, whether behind the doors and drawers of kitchen cabinetry or as a space unto itself, surrounded by three walls—with or without a door. The latter provides easy access to items without the inevitable dust and grease that can accumulate in a kitchen, and can be as large or small as you like.

The walk-in pantry featured here, for instance, is the size of a small room. Just steps away from the kitchen proper, it's spacious enough to accommodate not only provisions but also dinnerware, serveware, and food-prep space, complete with cutting boards. The homeowner took advantage of every inch of space, too. While drawers are devoted to linens, dishes and serving pieces find their places on shelves both high and low. And wherever possible, foodstuffs are stored in glass jars. For the remaining items, though, innovation takes over; stainless-steel turntables—typically used for spices—allow bottles and canned goods to be "stacked."

Turntables like these are typically reserved for spices. Here, though, they hold a wide variety of canned and jarred goods. Plus, the two-tiered styles make the most of these shelves' vertical space.

This antique epergne may well have originally displayed fruits or even desserts. But now it serves as a convenient tea caddy, complete with a wide variety of packets.

above • This wall rack provides a clever twist on organization, offering an easily accessible spot for cups and mugs.

facing page • A pantry adjacent to the kitchen keeps items close at hand without taking up valuable cabinet space in the kitchen proper. This one smartly has no door, so opening and closing it isn't an issue.

APPLIANCES

Appliances are the workhorses of any kitchen, performing tasks that range from keeping food cold to heating up meals—and washing the dishes afterward. Plus, new technology has made appliances smarter than ever, making kitchen duty easier, too.

Consider the Possibilities

Along with cabinetry, appliances account for the better part of a kitchen's budget. That said, this is no place to skimp. Because kitchen appliances are used on a daily basis, it's important that they can stand up to your family's needs. Start by asking yourself if the kitchen, for the most part, is used only by your immediate family; if so, a single range, refrigerator, and dishwasher should do. On the other hand, if two cooks are apt to use the kitchen at the same time—and entertaining is often on the agenda—you may want to think about doubling up on appliances. Finally, once you've addressed the basic necessities, consider today's specialty appliances; ranging from warming drawers to speed-cook ovens, wine refrigerators to built-in coffee systems, they're sure to make your kitchen even more efficient.

Before purchasing any appliance for a remodeling project, it's important to coordinate the specifications with existing cabinets. For new construction, coordinate before the cabinets are even built or purchased. Likewise, plumbing and electrical plans require some thought. Keep in mind, for instance, that steam ovens, coffee centers, and refrigerators with icemakers or water dispensers require water that tastes good—not just potable water—but a centralized water filter can handle them all.

Coordinating appliances with cabinetry is an important part of any remodel. In this kitchen, for instance, the cooktop's white hood matches the frame-and-panel doors on either side of it, making the hood all but disappear.

left · The streamlined look of this contemporary kitchen owes much to corralling several appliances on a single wall. The built-in refrigerator and wall ovens blend quietly into the black backdrop, as does the TV.

below · If yours is a family that grabs breakfast and snacks on the go, a designated spot away from the kitchen's primary work zone is a good idea. The countertop here provides the perfect workspace, while an undercounter microwave stands ready to warm up items.

Updating Appliances

Even if your remodeling plans involve no more than replacing appliances, you can make a big difference in your kitchen in terms of updating aesthetics, adding state-of-the-art features, and enhancing energy efficiency.

Some appliances are easy to replace. A standard 30-in. range, for example, will fit into the same space as an older range of the same size, as will a new Energy Star®-rated 24-in. dishwasher. Likewise, you're apt to find a new freestanding, counter-depth refrigerator that fits into your existing opening.

Where things become a bit more challenging, however, is when new built-in appliances need to fit into existing cut-outs. Most older American cabinetry is face-framed and—unless you know which company made them and can get specifications—there's no way to know the weight or cut-out tolerances allowed by the manufacturer. If you find yourself in this situation, it's worth paying a professional for their advice. And don't buy panel-ready appliances to fit into existing cabinetry; unless you are refinishing or refacing your cabinets, the two will be nearly impossible to match.

Freestanding vs. Built-In Appliances

One of the first decisions most homeowners make in terms of appliances is whether they want freestanding or built-in models. Like anything else, there are pros and cons to each. Freestanding appliances are less expensive and easier to install. On the downside, though, there are some restrictions; in a freestanding range, for example, the cooktop and oven must both be either gas or electric. There is no option to have one of each. Built-ins, on the other hand, allow the option of having a gas cooktop and an electric oven. Sometimes referred to as slide-ins, built-ins are typically flush to cabinets and countertops, creating a more streamlined look. Most are available in a variety of finishes, too, and have the option to add a panel or overlay to match your cabinetry. The disadvantage, however, is that this flexibility comes at a price; built-ins are costlier than their freestanding counterparts.

Freestanding range options include the traditional 30-in. model. Its disadvantage, though, is that the controls are on a raised back panel, which can be hard to see and reach. A slide-in range has the controls upfront, though one with the same features as the freestanding model can cost almost twice as much. Likewise, an alternative to a built-in refrigerator is a freestanding, countertop-depth model—easy to plan into a kitchen remodel. Be aware, however, that while it will fit into most existing openings, the doors and handles will extend beyond the cabinet faces. A standard-depth refrigerator, on the other hand, will extend into the kitchen, unless you recess the wall space to give it a cabinet-depth look from the front.

Many appliances—including dishwashers and warming drawers—can be paneled to match the surrounding cabinetry for an integrated look. Even built-in ovens and microwaves now offer sleeker frames that extend minimally, or not at all, from their cabinetry cutouts. But the cost for this kind of premium appliance package isn't cheap; it can be three to four times higher than what you'd pay for a standard suite.

left • In a kitchen with diminutive dimensions, it's important to keep appliances as low-profile as possible. A sleek cooktop and below-counter ovens do that here, while the overhead placement of the microwave keeps it up and out of the way. Even the counter-depth refrigerator does its part to remain low-key.

below and left • On one side of this kitchen, appliances are built in; the stainless-steel range, hood, and undercounter oven punctuate their otherwise all-white surroundings. On the opposite side of the room, however, space is at a minimum. Thus, the integrated refrigerator blends quietly into the backdrop.

This built-in range is flush to cabinets and countertops on either side, creating a clean, uncluttered look. In lieu of a conventional backsplash, a window is positioned between the range and hood, providing more light for the cooktop.

Integrated Appliances

When the kitchen is completely visible to adjacent living spaces, as in so many of today's open floor plans, homeowners often want appliances to all but disappear. As a result, integrated appliances—often in the form of dishwashers, warming drawers, and refrigerators—are more prevalent than ever, especially in contemporary, European-inspired spaces.

Unlike built-in appliances, the components of fully integrated appliances are seamlessly hidden behind cabinet panels. (This includes ventilation that's disguised by cabinetry.) An integrated refrigerator, for instance, might have the same frame and panel face as the rest of the kitchen's cabinetry.

Not surprisingly, it's important that they be professionally installed; custom panels, hinges, and/or hardware require technical precision in order to complete a smooth transformation. This high level of customization doesn't come without a hefty price tag, however, especially when compared to traditional appliances.

At first glance, the doors and drawers to the right of these wall ovens appear to be a part of a storage wall, their contemporary hardware emphasizing the linear grain of the wood. In fact, a refrigerator is behind the integrated façade, with freezer drawers just beneath it.

Dishwashers are a prime candidate to be integrated; here, the appliance is disguised by a steel-blue façade that matches the rest of the room's cabinetry. This integrated front is particularly deceptive because, instead of one flat panel, it has the appearance of three drawers.

An appliance wall in this contemporary space includes an integrated refrigerator and freezer column. Their easy-to-clean stainless-steel skins not only match the surrounds of the adjacent ovens and wine cooler but also the kitchen's countertop and hood.

Cooking Appliances

Deciding what cooking appliances are best for your kitchen should be based, for the most part, on three factors—space, style, and accessibility. From a space point of view, it's much easier to fit a 30-in. or 36-in. range into a small kitchen than it is to make room for separate cooktops and wall ovens. Plus, a 30-in. or 36-in. range, which typically combines a cooking surface and single or double ovens, can also offer some of the same dual-fuel, convection, and warming features that separate ovens often do. Keep in mind, however, that the benefits of a single appliance start to diminish in 48-in. or 60-in. models, as they can take up the same amount of space as separate cooktops and wall ovens.

In terms of style, a freestanding range or separate cooktops and ovens are a good fit for almost any room; it all comes down to your personal preference. The exception may be a kitchen with a streamlined, European-inspired look; in that case, a sleek induction cooktop—with ovens concealed behind sliding cabinet doors—will suit the style better than a range.

Finally, no matter what your age, give some thought to accessibility. In lieu of a standard range, for instance—with ovens set so low to the floor that pulling out heavy entrées can be a challenge—consider side-by-side ovens mounted at elbow height. Likewise, a counter-height cooktop with controls in the front functions perfectly well for many homeowners, but for easier wheelchair or seated use, consider one mounted at a lower table height.

above • A counter-height cooktop with controls in the front is a good choice for anyone, as you don't have to reach over the potentially hot surface to reach them.

left • On both induction and radiant cooktops, circles typically mark the location of the burners. There are, however, some models in which the black ceramic glass appears to have no markings at all.

above · In a single cooking zone, this kitchen has the best of all worlds—a range as well as specialty ovens. Grouping the appliances makes the most of a relatively small space, saving steps for the chef in the process.

left · If space allows, a separate cooktop and ovens can be a good choice for universal design. Not only is this cooktop at countertop height, but wall ovens are also installed so they open at countertop height or below.

COOKTOPS AND RANGES

When it comes to cooktops, gas and electric have long been the two basic options, but today the choices are more far-reaching; induction or radiant heat are increasingly popular. It can be nearly impossible to tell the difference between the latter two. An induction cooktop and one that features radiant heat often look identical, with sleek ceramic glass surfaces instead of exposed coil rings. There is a difference, however, in how they provide heat. An induction cooktop uses an electromagnetic field that heats the cookware rather than the surface itself, and heats up instantly. A radiant cooktop heats and cools down gradually, much like a traditional oven.

Cooktops come in different sizes, from compact 30-in. ranges to those that are 48 in. wide or larger. Modular cooktops are available, too, which allow customized setups. Components—including standard gas burners, a wok cooker, grills, induction burners, and steamers—are typically available in 12-in., 15-in., or 24-in. sizes, which can be combined to create a custom, large-scale cooktop or separated into independent cooking stations.

Induction Cooktops

Induction cooktops are powered by electromagnetic energy, causing cookware and the food in it to heat up while the surface itself remains cool to the touch. The burners are topped by a ceramic glass surface and, because they are quick to respond, induction cooktops are more energy efficient—by almost twice—than their gas or electric counterparts. Plus, they provide consistent heat and are easy to clean; spilled food won't burn and stick to the cooktop.

There are, however, drawbacks to consider. First, induction cooktops require flat-bottomed magnetic cookware, such as stainless steel. (The steel content of the cookware can be tested by taking a small magnet to the store.) Second, the price of an induction cooktop can be considerably more than a more conventional equivalent. Given that this is a good choice for universal design, however, the investment can be well worth it.

facing page · The wall ovens and cooktop are separate in this space, leaving room beneath the latter for storing pots and pans. Because these kitchen essentials are on a shelf and not in a drawer, they're even easier to access.

right · Side-swing doors aren't the only unique feature of this range; the cooktop goes beyond standard burners to accommodate your specific cooking needs. The color of this appliance makes it eye-catching on its own, but the chrome hood draws even more attention to it.

left · The thought of a commercial-style range often conjures up images of an enormous appliance. But this model scales it down to home-cook size, its white façade blending seamlessly with the cabinetry that surrounds it.

Cooking Appliances   99

A cooktop can be installed into a countertop like a drop-in sink—with the countertop surrounding it on all sides—or like an apron-front sink, with the countertop interrupted by the unit. Be sure that the backsplash is resistant to heat, and to moisture and stains, too. Just as important, allow at least 24 in. of space on either side of a cooktop, one side for food prep and the other a "landing space" for hot pots.

Like cooktops, today's ranges offer alternatives that reach far beyond gas and electric. There are dual-fuel options, for instance, even on the cooktop itself. And range configurations have changed to suit the times, with some moderately priced models featuring warming drawers or stacked ovens.

It's important to note that some ranges are taller and deeper than typical base cabinets and countertops. Likewise, a freestanding range may not fit perfectly into a spot where a slide-in once was. Compare your own situation to exact product specs and enquire about filler pieces and retrofit options. It's easier to accommodate bigger ranges in new cabinetry, as base cabinets can be installed an inch or two away from the wall (which will require deeper counter-tops) or on higher toe-space bases.

Dual-fuel ranges offer the best of both worlds, like this one with a pair of electric ovens and the easy-to-control heat of a gas cooktop. The components of this range are interchangeable, too; standard burners can be swapped out for a grill or griddle.

Sleek induction and radiant-heat cooktops are a good choice for today's contemporary kitchens. Their low-profile styling is typically just as slick and streamlined as the rest of the room.

Dual-Fuel Ranges

Serious bakers tend to prefer the precision of an electric oven; because it typically stays within 10° of the selected temperature, an electric oven ensures even cooking and keeps food from drying out. Serious cooks, on the other hand, appreciate the adjustable flame and versatility of a gas cooktop. Thanks to the dual-fuel range, a single appliance can accommodate both.

Originally, dual-fuel ranges were available only from high-end manufacturers in sizes that ran 36 in. or larger, with prices as substantial as their dimensions. Today, however—as sizes and prices have been scaled back—dual-fuel ranges are much more accessible to the average homeowner. There are plenty of less expensive, 30-in. models on the market. That's not to say that a dual-fuel range isn't without its downside, though; compared to a standard range of the same size, it's more expensive to purchase and to install.

above · The beauty of gas cooktops is that they offer precision heat control; some burners are occasionally dedicated, in fact, to simmer mode. Plus, adjacent burners make it easy to slide a pan from one to the next.

left · This freestanding range may have a vintage look, but it has all of the modern conveniences. The five-burner, dual-fuel model has convection capabilities as well as a warming drawer.

OVENS

The standard oven is a radiant, or thermal, oven, which cooks by a combination of radiant energy from a heat source; it may be an electric element at the bottom and top (for broiling) or a gas-fired flame beneath the oven floor. Convection ovens incorporate a fan in an electric radiant oven to speed up the cooking process, and a true convection oven has a third heating element that heats the air before it circulates, making it even more efficient.

If you tend to cook different foods at different temperatures, two ovens may serve you better than one. Wall ovens are available at all price points, and some 30-in. and 36-in. ranges offer two ovens, typically stacked. Larger pro-style ranges and heavy-duty European models, meanwhile, have several small ovens both stacked and side by side. On the other hand, a smaller oven is more energy efficient and typically retains moisture better, too.

Speed and Steam Ovens

The popularity of specialty ovens is on the rise, with speed and steam ovens among the most sought after. Speed ovens can incorporate microwave energy, steam, convection, and high-intensity radiant heat, such as halogen bulbs. Steam ovens incorporate steam from a reservoir or plumbed water line (the latter is more expensive), while the hybrid steam-convection oven provides the best of both technologies. You might, for instance, first steam a chicken to make it tender and moist, and then give it a crispy skin via convection. Either function can also be used alone.

left · Although the cooktop and oven are separate in this contemporary kitchen, they take up no more space than a conventional range. They do, however, provide the option of mixing and matching elements, giving a homeowner complete flexibility.

below · The compact nature of built-in steam and speed ovens makes them easy to combine with coffee systems. A warming drawer can easily be stacked beneath them, too, forming an efficient configuration.

above · Under an island's countertop is an ideal place for a steam, speed, or microwave oven. It's especially smart to keep the oven removed from the primary work space, so it can be used anytime—even while a meal is being prepared.

left · If you tend to cook at two different temperatures—or for a crowd—a range with a double-oven may best suit your lifestyle. This gas model features multiple burners, too, further increasing the range's capabilities.

Cooking Technology

There's no doubt about it. Cooking technology continues to improve and homeowners' lifestyles are easier because of it. There are more Internet-connected appliances than ever before, not to mention docking stations. Here are some of the latest innovations:

- French-door wall ovens are a relatively new offering in their own right; they're a smart choice in small spaces because you don't have to allow room in front for a conventional pull-down door. Some, though, are equipped with Bluetooth® so that you can control oven functions remotely with your smartphone—if you want to preheat the oven while driving home from the office, for instance.

- Induction cooktops are widely admired because they're quick to heat up, and now touch-and-swipe controls make them even more attractive. By dragging your finger around the arc of electronic controls—in a smartphone-inspired way—the heat can be raised or lowered in an instant.

- Some combination steam/convection ovens now come with a series of pre-set recipes. The oven does the work of figuring out which modes to apply; all you have to do is put in the food. You can even tell the oven what time you want your dish to be ready, and the oven will turn on—and adjust the cooking process as necessary—so it finishes at just the right moment.

With the light speed at which technology continues to advance, there's no telling what innovations will show up next.

This gas and dual-fuel slide-in range comes with six burners on a 30 in. model and a double oven—all in the same space used in a standard upright range. It also has a precision cooking probe that can communicate with a smartphone or tablet via Bluetooth (left).

Long gone are the days when your only choices were a conventional oven and a microwave. Today's plethora of new products—from speed and steam ovens to convection and combi-steam—allow you to customize a kitchen to your exact needs.

The microwave may not be a cutting-edge concept but its placement in the kitchen has become more innovative. In this kitchen, the appliance appears to be tucked under a countertop. In fact, though, the entire unit pulls out like a drawer, making it more accessible. Nearby, a warming drawer pulls out, too, its custom front matching the cabinetry.

COOKING VENTILATION

Good ventilation is essential to any cooktop. A ducted vent draws cooking smells, smoke, moisture, and airborne grease away from the cooktop, pulling them through the ducts and to the outdoors. A ductless hood vent, on the other hand, filters grease and moisture through a recirculation kit and then pushes the air back into the kitchen.

A built-in downdraft cooktop vent, which has a fan mounted under the cooktop, can pop up or be surface mounted. Pop-ups are better at venting than surface-mounted models, simply because they are physically closer to the tops of pots and pans. Both types of downdraft vents are less visible—and typically less expensive—than an updraft vent hood, but they may not be strong enough for high-intensity cooking. If an over-the-range microwave oven is your choice, select a ducted model and vent it to the outdoors. Finally, if your cooktop is in an island, be aware that the vent will require a stronger fan than if the cooktop were against a wall, as the wall helps direct heat, moisture, and grease to the hood.

A fan's strength—along with the hood's size, placement, and configuration—greatly influences the effectiveness of a ventilation system. A hood should be about 3 in. wider than the cooktop to scoop air most effectively, and even wider on an island cooktop. And the higher the hood is above the cooktop, the wider it should be.

Sometimes the simplest approach is best, exemplified by the hood in this kitchen. The boxy frame-and-panel piece blends quietly into the background, connecting to the ceiling via a cove molding.

above · The range in this kitchen slides neatly into an alcove, while—in keeping with the kitchen's overall style—the vent hides discreetly behind a built-in beam.

above · A pine hood is the undisputed focal point in this kitchen, going so far as to incorporate a ledge to display a favorite work of art. Make no mistake, though—it's but a veneer covering a standard vent, so safety is still the top priority.

right · A vent located over an island needs to have a stronger fan than if it were against a wall. This one has enough strength to pull heat, moisture, and grease, while the stainless-steel-and-glass piece has a sculptural feeling, too.

Cooking Appliances

RANGES

$ to $$$

- Combine a cooktop with one or two ovens below, creating a single 24-in. or larger appliance with four to eight burners and accessories. The most common size is 30 in.

- The cooking surface can be gas, electric, or induction, while ovens can be gas or electric. Dual-fuel models comprise gas cooktops and electric ovens.

- Cooktop accessories for 36-in. and larger ranges might include grills, griddles, and even wok burners.

- Ranges can include double ovens, convection ovens, or warming drawers, even in standard 30-in. models.

- Ranges are typically offered in freestanding models, with the controls on a raised panel behind the burners, or slide-in versions with controls in the front and no raised back panel.

- Stainless steel has long been the preferred finish for ranges, but glossy white, smartphone-inspired models are becoming more and more prevalent.

COOKTOPS

$ to $$$

- Cooktops can be set directly on a countertop with top-mounted controls or, in a range-top configuration, extend into the cabinetry below with front-mounted controls.

- The units can be gas, electric, or induction; modular units allow you to combine cooking types by pairing different burner modules.

- The most common sizes for electric and induction cooktops are 30 in. and 36 in. In gas models, 48-in. and 60-in. sizes are the most popular.

- Induction cooktops are the most energy-efficient and the easiest to clean.

above · Double-oven ranges are not at all uncommon today. Although they require extra width, these appliances can accommodate six to eight burners, some of which can be switched out for grills and griddles.

left · A generously sized cooktop requires ventilation that can handle the amount of heat, moisture, and steam it emits. The best solution in this contemporary kitchen was to install side-by-side hoods.

right · Combination wall ovens typically pair a standard and/or convection oven with a microwave, providing multiple cooking options in a compact space.

below · Cooktops can be set directly on a countertop with top-mounted controls or, in a range-top configuration—as in this kitchen—extended into the cabinetry below with front-mounted controls.

WALL OVENS

$$ TO $$$

- Wall ovens come in single or double versions. Single models are often mounted directly below a cooktop.

- For added practicality, combination wall ovens often pair a standard or convection oven with a convection microwave oven.

- Although referred to as wall ovens, single ovens are often mounted in a base cabinet, whereas double ovens are usually mounted in tall cabinets.

- Speed- and steam-cooking features are increasingly popular in wall ovens. Some models may also include a pizza mode, designed to produce the extra heat associated with brick ovens.

MICROWAVE OVENS

$ TO $$$

- Microwave ovens are available in countertop models, built-in models with trim kits, over-the-range models with vent fans, and drawer models that mount below countertops.

- Many of today's microwaves offer multitasking functions such as speed-cooking, convection, and warming capabilities.

- No longer an appliance primarily intended for warming up food, some microwaves can broil, roast, and bake.

- Pre-programmed settings and sensor heating take the guesswork out of cooking with microwaves.

Refrigerators and Freezers

Refrigerators and freezers have gone beyond the once-standard, twin-size-bed appliances; while they're still available, other options are extensive. You'll find models in various configurations, their interiors fitted with drawers, moveable shelves, variable temperatures, and electric controls. And specialized storage zones are now the norm, including temperature-controlled drawers for thawing frozen meats and freezer pockets for boxed pizzas.

Freestanding refrigerators come in standard- and counter-depth models. A standard-depth refrigerator is typically about 68 in. to 71 in. tall and 31 in. to 35 in. deep to the door fronts. Available in side-by-side, bottom-freezer, and top-freezer options, it's an affordable choice that provides the most cubic ft. of storage in the least expanse of height and of width. Meanwhile, as its name implies, a counter-depth refrigerator is only as deep as your countertops. Typically, 32 in. to 36 in. wide, 24 in. to 26 in. deep (not including the door), and 68 in. to 71 in. tall, it offers the look of a built-in refrigerator with one distinct advantage: It's often thousands of dollars less than a true built-in refrigerator.

True built-in refrigerators are flush (including the door) with the depth of most standard base cabinets; you'll find stainless-steel models as well as some with panel-ready and glass-front options. Even with matching cabinet panels, however, the flange along the outside of the unit and the grille on top will be visible. If you want it to be completely hidden from view, a fully integrated refrigerator is the solution.

You will find options beyond the single upright unit, too. Both refrigerators and freezers are now available in independent columns that can be installed side by side or in separate locations of the kitchen. Plus, refrigerator drawers can be installed below countertops, providing supplemental storage.

This integrated refrigerator/freezer is fronted by the same paneling used throughout the rest of the room. Its only identifying factors are the door pulls and trim above and below the grille.

An impressive cupboard in this kitchen isn't a cabinet at all. The doors open to reveal refrigeration units, while the drawers are dedicated to freezer space.

Built-in refrigerators, like the one in this kitchen, are flush with the depth of most standard base cabinets. Stainless-steel models have long been the most prevalent, but panel-ready and glass-front options continue to grow in popularity.

Cooling Appliances

COMBINATION REFRIGERATOR/FREEZERS

$ to $$$

- Freestanding refrigerator/freezers come in built-in or freestanding models.

- These appliances can be standard depth (typically between 31 in. and 34 in. with handles) or counter depth, in which only the doors and handles extend past 24-in.-deep cabinet fronts.

- Configurations include top-mount freezer, bottom-mount freezer, side-by-side, and French door with single or double drawers.

- Some of today's many options include adjustable bins and shelves, in-door ice makers, and Wi-Fi that lets you access apps, leaves notes, view recipes, and more.

REFRIGERATOR AND FREEZER COLUMNS

$$$

- All-refrigerator and all-freezer columns can be placed next to one another or in separate locations.

- Available in stainless-steel or panel-ready options, these offerings allow for complete customization, according to a homeowner's food-storage and space needs.

- Typical widths are 18 in., 24 in., 30 in., and 36 in.; heights range from 80 in. to 84 in.

- One of the most popular uses for refrigerator columns is for wine storage.

above • Some refrigerator/freezer options combine solid and see-through doors, the latter making it easy to see at a glance what needs to be put on the grocery list.

right • The outward appearance of this drawer makes it look like the rest of the kitchen's cabinetry. It opens, however, to expose freezer space, handy for the family's favorite snacks.

right · A wine cooler provides climate-controlled storage. This undercounter model fits easily into the kitchen and—placed at one end of the island—is easily accessible from the nearby eating area.

below · Stainless-steel refrigerators are still the standard for many a kitchen. The French doors of this model open to reveal the refrigeration unit, while the drawer below provides easily accessible freezer space.

REFRIGERATOR AND FREEZER DRAWERS
$$ to $$$

- Refrigerator and freezer drawers have become increasingly popular as supplemental, cold-food storage.

- Drawers are available in 15-in., 24-in., 27-in., 30-in., and 36-in. widths.

- Units are available in two-drawer configurations, with all-refrigerator, all-freezer, refrigerator/freezer, and ice maker options.

- You'll find these appliances in popular cabinet finishes or panel-ready.

WINE AND BEVERAGE REFRIGERATION
$ to $$$

- Wine refrigeration provides climate-controlled storage of a homeowner's collection.

- Full-height columns are typically 30 in. wide, while undercounter versions are 15 in., 18 in., or 24 in. wide.

- Units are available in single- and double-zone models; the latter controls the temperature of both red and white wines.

- Some beverage refrigerators are devoted to soda and water storage—with a small bin for snacks—while others have a place for wine, too.

Dishwashers

Like every other kitchen appliance, dishwashers have seen a number of style and technology advancements, giving the homeowner more options than ever. The standard dishwasher is 24 in. wide, although 18-in. models are available for tight spaces; extra-wide dishwashers are available, too. When deciding what size is best for you, take some of your most-used dishes to the showroom to test a variety of rack configurations. Some dishwashers have racks that can move down, making room for plates on the top, while others have tines that fold out of the way to make room for pots and pans. You'll find those that even offer a steam option, which is a gentle way to wash fragile stemware. And if you want a little extra space—or are particular about silverware—look for a dishwasher with a third rack. You'll find a variety of settings, too, ranging from those that will delicately clean china and crystal to those that will scrub pots and pans.

And the options don't stop there. Timers allow you to schedule a load when the household is asleep or during off-peak energy hours, and integrated consoles keep controls hidden from view. App-enabled models will even let you check on the progress of your cycle while you're away from home.

Traditional dishwashers are still the most popular option, available not only in stainless-steel and panel-ready styles but also in slate and black stainless finishes. But dishwasher drawers are quickly gaining on them. Because they're intended for smaller loads, dishwasher drawers—available in single- and double-drawer configurations—are more efficient, too. And, like their conventional counterparts, they come in a variety of finishes as well as panel-ready models.

This dishwasher is the same size as a conventional model but, instead of a pull-down door, it pulls out like a drawer, making it more easily accessible.

In kitchens with integrated dishwashers, the appliance can be hard to detect; the front is paneled to match the rest of the room's cabinetry.

If you entertain often, you may want a dishwasher drawer to supplement a standard model. You might even look for a drawer with a steam option, a good way to clean fragile glasses and stemware.

The Little Extras

When considering what you need to make your kitchen its most efficient, think beyond the obvious large appliances. Here are just a few of the other options you may want to consider.

COFFEE SYSTEMS

Built-in coffee systems offer an upscale alternative to countertop machines. Some models come with plumbing hookups, while others rely on reservoir systems. Although they come with a higher price tag, it may be worth it to the coffee connoisseur to enjoy their favorite coffee drink in a matter of minutes. For an added element of luxury, a cup-warming drawer can be located beneath the machine. At the same time, however, you can put together a custom coffee bar with a combination of small appliances.

WARMING DRAWERS

A warming drawer can perform more functions than you may think. While it does, in fact, keep prepared meals hot, it can also warm plates, slow-cook, and even proof bread dough. Typically, a warming drawer is installed below a wall oven, but it can also be placed below a countertop for greater convenience. Available in stainless-steel or panel-ready options, warming drawers can easily coordinate with the rest of your kitchen, too.

above • Located just around the corner from the kitchen proper, a corner of this pantry is dedicated to a coffee bar. There's no need for a high-end coffee system here—a standard coffee pot and compact espresso machine fill the bill.

left • Whether your preference is coffee, cappuccino, or espresso, a built-in coffee system like this one is at your service. Many of today's machines make two drinks at one time, and clean the milk container after each use, too.

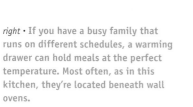

right · If you have a busy family that runs on different schedules, a warming drawer can hold meals at the perfect temperature. Most often, as in this kitchen, they're located beneath wall ovens.

below · Pizza ovens are by no means reserved for a teen's favorite treat; they can bake everything from bread to chicken. A pizza oven does, however, require a supportive concrete base, which should be installed by a professional mason.

PIZZA OVENS

Pizza ovens are somewhat of a misnomer; not only do they turn out perfectly baked pizza pies but also everything from bread to chicken—making them bonus cooking zones. They do require careful planning, however. A pizza oven requires a concrete base to support the weight of the stone around the oven, and the oven's interior should be made of clay-fired bricks to stand up to high temperatures and hold the heat well. The exterior, though, can be made of any fireproof material. Although it takes some time for the oven to heat up, once it's reached its optimal temperature food actually cooks faster.

ICE MAKERS

If you entertain on a regular basis, a dedicated ice maker might be on your list of must-haves. It makes ice cubes using distilled or well-filtered water, creating perfectly clear cubes with no trapped air. Ice makers come in a variety of sizes and configurations, ranging from 15 in. to 24 in. wide. in undercounter, built-in, freestanding, and countertop styles. While the majority of ice makers are stainless steel, high-end models are available in panel-ready styles.

TRASH COMPACTORS

The installation of a trash compactor can greatly reduce the number of trips to the curbside trash bin—as well as what you send to landfills. Most compactors are available in 12-in. to 15-in. widths, with heights of approximately 34 in. or 35 in. Look for one with a toe bar; when your arms are full of garbage to throw away, this simple feature will make your life easier. Just as importantly, the trash compactor you choose should include air freshener compartments or charcoal filters to reduce odors.

Incorporating Electronics

There's nary a kitchen today that doesn't include some type of electronics, in the form of televisions, computers, or even tablets. As with any conventional appliance, the incorporation of these items requires some forethought.

TELEVISIONS

While televisions have been at home in the kitchen for quite some time, advanced technology—not to mention larger models—has made them more of a challenge to incorporate. Because most require a cable box or satellite dish, their place in the kitchen can be limited. (It's ideal if a remote location is available for a cable box.) There are a number of placement options, however, some easier to achieve than others. When a TV is incorporated into an appliance wall, for instance, it blends in quietly with the rest of the room's streamlined look. Likewise, a television can do a disappearing act; drop-down styles can be tucked beneath wall cabinets, while pop-up models might be placed at the end of an island. You can even place a small TV on a shelf or tuck it into a niche. Just be sure that it's at the proper viewing level and has the right vantage point. Will it be primarily intended for the cook or for those seated at an island or breakfast nook?

A TV is central to almost any great room, and this one is no exception. The screen, set into a perfectly-sized niche, is visible from the dining area as well as the adjacent kitchen.

A small TV hides behind cabinet doors at the end of this short peninsula. When a family member wants to catch the news or watch a favorite show, the doors slide back and out of the way, allowing the TV to be pivoted and seen from several angles.

COMPUTERS AND TABLETS

There was a time when home offices were incorporated into the kitchen primarily to provide a place for computers, making paying bills or doing homework more convenient. But today computers play a much larger role. They're handy for sending emails, searching out recipes, watching the news, even keeping track of the family's schedules. Plus, desktop computers have, for the most part, given way to notebooks and iPads—which require less space. With these devices, nearly half of today's remodelers are incorporating smart technology, in the form of security, entertainment, climate control, and/or lighting.

There are safety concerns with bringing portable electronics into the kitchen, however. Try to keep ingredients at a safe distance and take care not to use them near water. Today's electronics aren't as delicate as they once were, but neither are they waterproof.

above • A quick glance at this TV might make it appear to be another oven stacked over the microwave. It's not only subtle but also smartly placed well out of the way of splatters but easy to watch at the same time.

right • Because laptops and tablets are frequently used in today's kitchen, it's crucial to designate a space away from heat and water for their use. A power source is essential, too, handled here by outlets tucked under the countertop.

Efficiency at Its Best

When putting together a kitchen design, efficiency should have a high priority. Because appliances can be the most expensive part of a kitchen remodel, deciding which one will go where is an important—and ultimately step-saving—decision.

For starters, consider the tried-and-true work triangle, which contends that the stove, sink, and refrigerator should be within easy reach of each other. As kitchens continue to grow in size, however, it's more feasible that a kitchen's three work centers—prep, cooking, and clean-up—simply have direct access to each other. You shouldn't have to walk around an island, for example, to get from one to another.

In large kitchens like the one shown here, the conventional work triangle often gives way to specific zones, instead. A prep sink might be conveniently located next to the refrigerator, or a microwave may be close to the main sink. In sizeable kitchens, also consider keeping primary appliances close to the room's core, placing lesser-used appliances at the perimeter.

Finally, don't forget that most kitchens aren't reserved only for cooking. Take into account the social aspect by keeping guests' pathways from crisscrossing the chef's work zones.

above • The main appliances in this kitchen adhere to the conventional work triangle; the refrigerator, cooktop, and primary sink are just a few steps away from one another. Meanwhile, the microwave and wall ovens—as well as a prep sink—are smartly placed at the kitchen's edge.

right • Double ovens and a warming drawer are placed at the edge of the kitchen, with a landing area to their immediate right. The location of the appliances is convenient yet out of the kitchen's main traffic pattern.

left · A refrigerator requires a nearby sink but this kitchen goes a step further. The main sink, set in the island, is just a few steps away, while a prep sink in the far counter isn't much farther.

right · A storage wall to the left of the pantry door incorporates a microwave as well as a TV. The microwave is set high enough to make it easy to reach into and far enough away from the kitchen's primary triangle that a user won't get in the way of the chef.

COUNTERTOPS

Countertops and backsplashes are the hardworking superstars of any kitchen,

AND

with the ability to be just as fashionable as they are functional.

BACKSPLASHES

Countertops

The ideal countertop has it all. It's nonporous, non-staining, heatproof, durable, easy to clean, scratch resistant—and good looking. And, in a perfect world, inexpensive, too. In reality, though, you'll probably need to compromise here and there to find the countertop that best suits your needs. The good news is that there is an ever-growing selection of counter-tops from which to choose, from increasingly attractive laminate to natural and engineered stone, easy-to-maintain solid surfacing to stainless steel, wood, bamboo, and even recycled paper.

When choosing countertops, consider how and where they'll be used. Will you use the same material throughout the kitchen or combine two or three to create zones for specific purposes? Perhaps you want to devote all or some of your countertops to natural stone. Marble is unquestionably stunning—and its naturally cool temperature makes for great pastry making—but will you still love it after years of use? Marble stains and shows wear over time, although some homeowners like its worn patina. Additionally, upkeep should be considered. For instance, the idea of a butcher-block prep counter might be appealing, but keep in mind you'll have to oil it on a regular basis.

Stone countertops can add an element of luxury, especially when the kitchen opens onto a living area. In this case, one space is just as upscale as the other.

Wood adds an element of warmth to any kitchen, exemplified by the wood-topped island in this traditional kitchen. The work portion of this island is not to be confused with a chopping block, however; it's important to use cutting boards at all times.

Limestone countertops are a perfect fit for this contemporary kitchen, the pale gray complementing the room's stainless-steel and chrome elements.

There's a multitude of look-alike countertop materials available today, making it easy to find one that will work for your needs—and your price point, too. This island, for example, has the upscale look of marble, but a hard-wearing quartz could present the same appearance.

Mixing Countertop Materials

Countertops can easily be mixed and matched to suit different kitchen tasks. A marble counter, for example, provides a cool, hard surface for rolling out dough, while stainless steel makes a sanitary, heatproof prep zone.

Mixing countertop materials offers more than functional benefits, however. You can save money by using a less expensive material for perimeter countertops while splurging on something more luxurious for the island. In addition, mixing materials adds visual interest to a kitchen, keeping any one color or material from being used excessively. The possibilities are all but endless. Pair marble with walnut, granite with butcher block, even quartz with stainless steel; go wherever your imagination leads you.

Topped with marble, the island of this kitchen gets a luxurious touch, while wooden perimeter counters are more hard-working. Wood countertops like these need to be specially treated; around the sink, for instance, the wood needs to be waterproof.

Butcher block tops the island in this clean-lined kitchen, providing a durable surface for chopping vegetables at one end and dining at the other. Meanwhile, perimeter countertops are made of stainless steel, a material so hard-wearing it's the choice of professional chefs.

above • The perimeter cabinets of this kitchen are topped with quartz; both heat-and-scratch resistant, it's one of the toughest countertops you'll find. Meanwhile, the island showcases a stunning slab of marble.

right • A mix of countertop materials is utilized in this kitchen, the type dictated by their purpose. Chosen for its durability and ease of maintenance, white laminate defines the sink and cooktop areas. A section of wood is added, though, for the expanse dedicated to dining.

NATURAL STONE

The natural appeal of slab stone is undeniable, with marble, granite, and soapstone among the most popular. But because stone will take a large chunk out of your remodeling budget, it's important to know the pros and cons of each.

Marble

Color variations that occur in marble are due to impurities in the stone; that's why slabs—as opposed to tiles—are used for countertops, where the dramatic veining can be showcased. In addition to those seductive good looks, marble is favored for the soft patina it develops over time. Maintenance, though, can be a factor. While polished marble isn't likely to stain if properly sealed, acidic foods will etch the surface, even if spills are wiped up right away. Honed marble, on the other hand, doesn't show scratches as much as its polished counterpart.

Granite

Granite has a lot going for it. In addition to its good looks—be it mottled, monochromatic, or veined— this natural stone is extremely hard, making it tough to scratch. Plus, it's cool to the touch and can withstand heat well. On the downside, it can be hard on dishes and can stain unless it's sealed regularly. There are environmental aspects to be considered as well. Quarried in many countries—in a wide variety of colors, patterns, and prices—granite can be environmentally friendly, or not. Some granites may be available locally, whereas others might be quarried in, say, South America and travel the ocean to another continent before making their way to your local stone yard.

An island with waterfall edges is one of the most stylish ways to showcase a slab of marble. And the clean lines are a good fit for a contemporary kitchen.

Choosing Stone

When choosing a stone countertop, it's important to visit stone yards in person; because no two pieces are alike, you'll need to choose the exact slabs you want. Find out, too, how the slabs will be seamed. Vein-matched seams are made from two sequential slabs polished on the same face, whereas bookmarked seams are made from two sequential slabs polished on opposite faces so that they mirror each other when seamed.

Soapstone

Soapstone's durability has made it a go-to countertop for hundreds of years, its density making it resistant to stains and bacteria, and unaffected by heat. While it comes in smaller slabs than granite—and in more limited colors—soapstone has an advantage in that it typically doesn't need to be sealed. The stone is light in tone when quarried and polished, but oxidizes over time and turns dark, and veining becomes more prominent. Some homeowners opt to speed up the darkening process by rubbing soapstone with mineral oil, boiled linseed oil, or a food-safe beeswax, while others prefer to regularly sand the surface to keep it looking like new.

The density of soapstone makes it resistant to stains and bacteria, and therefore a good fit for any kitchen. Additionally, it has the edge over granite in terms of maintenance because it doesn't typically need to be sealed.

above • Luxurious marble makes its presence known throughout this traditional kitchen, used not only on the countertops but on the backsplashes, as well. The natural color variations of marble often inspire a room's color scheme, as in this gray-and-white space.

left • Because no two slabs of granite are alike—each has its own unique pattern and color combinations—it's important to pick your granite in person.

ENGINEERED STONE

Engineered stone is a man-made product formed from 90% to 95% ground quartz and 5% to 10% resins and pigments—and the marriage of the two materials is a good one. The resulting product is heat and scratch resistant like granite, but nonporous, less prone to scratches and stains, and practically maintenance-free. On the downside, however, it isn't heat tolerant.

Engineered quartz is also the chameleon of countertops, available in everything from monochromatic tones to patterns that mimic granite or marble. Whether your preference is modern or traditional, honed or polished, flecked or mottled, this material's myriad options have you covered. Sold under such brand names as Silestone®, Cambria®, Zodiaq®, and Caesarstone®, some offerings have recycled content and sustainability benefits, too.

above • Black-and-white kitchens never go out of style, but this one is as functional as it is fashionable. Hard-wearing black Caesarstone tops the white cabinetry; as a bonus, the dark color disguises any scratches or spills.

right • If all-white cabinets leave your kitchen looking a bit sterile, consider adding a quartz "skin." It's a good way to customize home center stock cabinetry—and add a pop of color at the same time.

Composites

Composite countertops are made up of wide-ranging materials; recycled metals, paper, and glass—even plant pulps, such as wheat paste and sorghum—are among those you'll find. Some of today's most popular options include the following:

- **Paper-based composites.** Made with recycled paper, newspaper, or cardboard mixed with resin, paper-based countertops have the look and feel of soapstone but at a fraction of the cost. They are stain resistant, food safe, and strong, although most require finishing with wax or mineral oil.

- **Recycled-glass composites.** Fabricated from recycled glass mixed with cement binder, these countertops have a makeup similar to terrazzo. Because both glass and cement are durable, together they make an almost indestructible duo. This composite countertop is heat resistant. Because the surface is porous, however, it does require occasional sealing.

- **Scrap-metal composites.** Typically made with aluminum shavings bound with resin, these countertops are installed on a plywood substrate. Scrap-metal composite counters don't require a sealer, but on the downside, aren't scratch or heat resistant.

- **EcoTop® composites.** Made of 50% bamboo and 50% recycled paper, EcoTop composites are fused with water-based resins. The nonporous material is approximately five times stronger than granite, doesn't absorb stains, and is scratch resistant. Available in a wide range of colors, it can be customized with edge detailing, too.

A solid-surface composite, PaperStone is an environmentally friendly material made from either post-consumer recycled paper or virgin paper and resins derived from cashew nut shells.

above · Quartz is one of the most durable countertops you will find, making Silestone a good choice for the island in this kitchen. The only downside of quartz is that it's not heat resistant; here, though, honed black granite—which handles heat well—surrounds the cooking zone and perimeter countertop.

left · Because it's available in everything from solid colors to patterns that have the look of natural stone, quartz is right at home in almost any kitchen. This example of Cambria blends beautifully with the perimeter cabinets and provides a welcome contrast to the wood-based island.

SOLID SURFACE AND LAMINATE

Solid-surface and laminate countertops have very different attributes. But they have something important in common, too: They're both extremely easy to care for.

Solid Surface

It's hard to tell at a glance whether solid-surface countertops are just that, or whether they're stone, wood, or plastic laminate; like engineered stone, they have the ability to imitate almost any material. Their makeup, however, is quite different. Solid-surface counters are a blend of acrylic or polyester resins with powdered fillers and pigments; cast into slabs, they can be formed into any configuration.

Solid surfacing is vulnerable to scratches, but its unique composition makes buffing them out a breeze. It is sensitive to heat, as well, but those are the only real downsides. Sold by manufacturers such as Wilsonart®, Formica®, and Dupont Corian®, solid surfacing has much going for it; the material is durable and nonporous, making it resistant to stains, mildew, and bacteria.

Laminate

In addition to being the most affordable countertop option, plastic laminate has a number of other advantages. It's not only stainproof and easy to clean, but a wide range of colors and patterns means it can be right at home anywhere, too. And to take customization one step further, you can choose among various edges to trim your countertop, such as profiled edges (beveled, ogee, or bullnose), plastic-edge bands, and metal edges. Keep in mind, however, that because laminate is a thin surface, it's not the most durable of countertops; sharp knives and hot pans can wreak havoc on an otherwise pristine surface.

Porcelain Slab

Just when you thought you had seen all that porcelain had to offer, innovative slab surfaces have emerged from Europe, providing new possibilities for kitchen counters. These nonporous materials are easy to maintain and, like granite and quartz, are heat and scratch resistant. And they are stronger than granite (meaning fewer chips and cracks) but lighter in weight, making them easier to install.

In fact, countertops made from porcelain offer a multitude of benefits. Not only is this material available in a wide range of colors; it comes in a variety of patterns, too, including those that mimic marble and other natural stones. In addition, slabs up to 10 ft. by 5 ft. are available, meaning fewer seams to break up the run of a countertop. Perhaps best of all, though, porcelain countertops can sometimes be installed over existing counters, reducing the need to demolish the old material before installing the new.

above • For all its stunning good looks, this kitchen doesn't give up a thing in terms of being hard-working. Island cabinetry is wrapped with Caesarstone, which provides eye-pleasing contrast. A solid surface material could have been used for the same effect. The color is a smart choice because a countertop that matched the dark storage drawers would have weighed the space down visually.

right • You can choose among various edges to trim a laminate countertop, such as profiled edges (beveled, ogee, or bullnose), plastic-edge bands, and metal edges. This contemporary countertop, however, proves that sometimes the best approach is the simplest.

above • Corian is stainproof and easy to clean, making it an equally good choice for kitchen counters and eating bars. The vast range of colors and patterns available paves the way for mixing and matching, too.

WOOD

Wood countertops add an element of warmth to a kitchen. Due to cost, availability, and maintenance, however, they are typically used in specific areas as opposed to the entire space—a functional butcher-block prep section or a decorative surface.

Strong woods make the best butcher-block counter-tops, with maple being one of the best options, although oak, walnut, and cherry are also possibilities. Both end-grain and edge-grain butcher block make excellent cutting surfaces; edge-grain is not quite as hard as end-grain, however, so it's a bit more susceptible to dents.

Work-surface wood countertops are typically unsealed, although they need to be oiled regularly with food-safe mineral, tung, or linseed oil. Wood surfaces in water-prone areas also need to be sealed. The species of wood you choose makes a big difference: Redwood, yellow cedar, mahogany, white oak, and teak are all naturally more resistant to water. Keep in mind, too, that butcher-block countertops can survive heat to a certain extent, but they can be scorched or stained by hot or wet cast-iron pans.

Decorative wood countertops—often designated for islands or eating bars—are generally crafted from matching planks, resulting in seamless runs with decorative edges. Walnut and cherry are popular options, but you'll also find bamboo and reclaimed woods. Like their butcher-block counterparts, decorative wood tops should be kept as dry as possible and properly maintained.

Bamboo

Bamboo is technically a grass, but it's harder than most wood species and it matures much more quickly; it's ready to harvest in seven years and grows back easily. That said, a considerable amount of adhesive is required to laminate bamboo strands together, so look for low- or no-volatile organic compound (VOC) glues. Bamboo can be laminated to show end grain, edge grain, or flat grain. It should be treated like wood, too, with an oil applied to work surfaces or polyurethane applied to all sides.

above • Although crafted of edge-grain pieces—which can make an excellent cutting surface—this wood countertop is reserved for dining. The shapely form is a nice change of pace, too, from the kitchen's otherwise linear look.

facing page • Often designated for islands, decorative wood countertops are generally crafted from matching planks, resulting in seamless runs. If you want to keep maintenance to a minimum, consider using reclaimed wood, which is already nicked and scratched.

While most islands are built in, this one is a freestanding piece that has the appearance of furniture. The handsome butcher-block top—a near match for the cabinetry—does its part, too, to create an upscale look.

The entire island in this kitchen is topped with edge-grain butcher block, creating an expansive work surface. Maple, as shown here, is one of the best options for butcher block, but oak, walnut, and cherry can also be used.

METAL

Stainless steel is by far the most popular metal countertop material. It's nonporous and nonstaining, heat resistant, and easy to clean. Even better, stainless steel can be formed with an integral sink or backsplash, making the countertop watertight. Keep in mind, however, that there's no way to avoid scratches; you might opt for a brushed surface that will better disguise fingerprints and scratches, though you'll also find satin (smooth), antique matte, and specialty patterns.

A 14- or 16-gauge stainless-steel countertop will suit most kitchens (the lower the gauge number, the thicker the steel), formed around or supported by plywood or MDF to add strength and mute sound. When budgeting for stainless, factor length into the price; 10 ft. is the cutoff before the price per square foot goes up. Some homeowners prefer the modified marine edge, a rounded lip that keeps smaller spills on the countertop, while others like how the square profile makes it easier to sweep chopped food into a pot. A bullnose edge is best for countertops that won't see water, as the half-round profile will allow dripping onto the cabinets below.

Still, other metals have their devotees. *Zinc* is softer than stainless steel and more moderately priced. It's nonporous, bacteria resistant, and needs only a quick wipe-down to be cleaned; plus, it develops a striking patina over time. Although expensive, *copper* will develop a patina, too—either naturally or with the application of chemicals or heat. Like zinc, however, it's soft and thus susceptible to scratches. Be careful, too, about putting hot cast-iron pans on zinc or copper as discoloration can occur. *Pewter* is extremely expensive and easy to scratch, but adds a European-kitchen look. *Bronze* oxidizes like copper but to a lesser degree; it's stronger than copper and somewhat more expensive.

above • Because pewter is one of the more expensive countertop options, you can stretch your budget by using it selectively. In this kitchen, for instance, it's the star of the show atop the central island.

right • Part of the beauty of stainless steel is that it can be paired with easy-to-clean integral sinks. This example goes a step further, though, incorporating an area to drain dishes.

Concrete

If you're considering concrete countertops, don't think for a minute that your options are limited to gray. Pigments, stains, and dyes can create concrete counters in a vast array of colors. And any number of objects can be cast into the material, ranging from glass bits to shells. The price you'll pay can vary greatly, as well, depending on what aggregates or additives are in the concrete.

Concrete countertops can be precast or cast in place; precast counters are cast offsite by a local artisan, while cast-in-place counters are poured right on top of cabinets and finished in place. (Because concrete is just as heavy as stone, account for the extra weight when choosing your cabinets.) This material is both heat resistant and durable, but because it stains, concrete needs to be treated periodically with a penetrating or topical sealer. Penetrating sealers are barely discernible once dry, though spills can leave lasting stains or marks more easily than on surfaces finished with topical sealers. Conversely, topical sealers—such as epoxy, urethane, and wax—vary in both their appearance and performance. Epoxy and urethane sealers are thick and glossy, whereas wax, although handsome and easy for the do-it-yourselfer to apply, doesn't perform quite as well.

The island of this kitchen features a honed soapstone countertop, which can take on the look of concrete. It's as durable as any countertop material you'll find, although small hairline cracks may develop over time.

left · While some kitchens use stainless steel in limited roles, others take an opposite tack. This kitchen, for instance, gets its cohesive appearance by featuring the metal not only on the countertops and cooktop hood but also on the appliances and barstools.

Countertops

NATURAL STONE

$$ to $$$$

- Most often granite or marble slab, but soapstone is also in this category
- Granite is typically polished to a reflective sheen or honed for a matte look
- Natural veining, pattern, and irregularities are part of the natural appeal
- Stone needs to be sealed periodically
- Granite is scratch and heat resistant but can be scorched and can dull knives
- Marble, a soft stone, is prone to chipping and staining

ENGINEERED STONE/QUARTZ

$$$

- Comparable in heat and scratch resistance to granite
- Some brands offer integral-style sinks
- Nonporous, so they never need to be sealed
- Available in solid colors and patterns that mimic natural stone
- Available in polished or matte finishes

SOLID SURFACE

$$

- Large selection of colors and patterns available
- Nearly seamless in appearance
- Most lines offer integral sinks and seamless coved backsplashes
- Requires no sealing
- Repairable

LAMINATE

$

- Most affordable countertop option to buy and install
- Simple layouts are best for do-it-yourself installations
- Widely available in home centers
- Works best with drop-in sinks vs. undermount models
- Easy to damage and generally not repairable

Both natural and engineered stone (quartz) countertops are available in polished or matte finishes, giving you the option of completely different looks.

Today's engineered-stone countertops imitate natural stone better than ever while providing superior stain resistance at the same time.

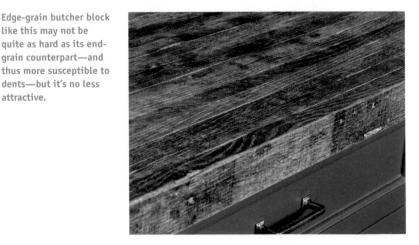

Edge-grain butcher block like this may not be quite as hard as its end-grain counterpart—and thus more susceptible to dents—but it's no less attractive.

The beauty of Corian in a kitchen is several fold: The solid surface material is tough, durable, repairable, and easy to clean.

PORCELAIN SLAB
$$$
- Extremely stain, heat, and scratch resistant
- Much thinner and lighter than stone, with comparable durability
- Can be installed over other counterparts for reduced replacement cost
- Most often shown with undermount sink, but integral sinks are available

WOOD
$$$
- Options include elegant teak or cherry planks or hardworking maple or oak butcher block
- Needs to be sealed against moisture
- Higher-end materials are used for focal points like islands or eating bars
- Can be sanded for scratch and stain removal
- Eco-friendly alternatives, such as reclaimed lumber and sustainably harvested species, are available

STAINLESS STEEL
$$$
- Hygienic and easy to clean but will show streaks, fingerprints, and water spots
- Often paired with an integral sink
- Good heat resistance
- Readily shows scratches, dents, and fingerprints
- Can be noisy

CONCRETE
$ to $$$
- Can be customized in a wide variety of colors, patterns, and textures
- Poured-in-place concrete is ideal for seamless appearance in oversize spaces or those interrupted by posts
- Can accept integral sinks
- Can crack and stain
- Needs to be sealed with protective finish
- Thicker installations require extra support

Backsplashes

Once a purely practical feature, the backsplash has come a long way. While it still protects kitchen walls from stains and splashes, a backsplash—more often than not—is equal parts fashion and function. Today's vast array of materials, in just as many finishes, means that it's easy to make a style statement, too. From ceramic and glass tile to stainless steel and beadboard, there's a backsplash to suit every kitchen design.

That's not to say that the practical aspect of a backsplash has any less importance. The perfect backsplash depends a great deal on where it will be placed. If it backs a food prep area or sink, a backsplash needs to be stain and/or water resistant. Likewise, if it's located behind a cooktop, a backsplash needs to stand up to a certain amount of heat. But if it's not adjacent to a workspace, a backsplash can beautifully showcase the material—or materials—of your choice. You're limited only by your imagination.

White subway tile is more than a mere backsplash in this kitchen; because it reaches all the way to the ceiling it's also a wall treatment, creating a textured backdrop for the bracket-supported shelves.

above • In a kitchen with otherwise pristine white walls, decorative tile can create an impressive focal point. Here, for instance, patterned tiles embellish the cooktop wall, at the same time providing a heat-resistant backsplash.

left • Tiles that have the appearance of wood offer the best of two worlds: the look of the real thing without all the maintenance.

The same marble that flanks this range runs right up the wall to create a matching backsplash. Full-height backsplashes like this create continuity but are typically more expensive installations.

BACKSPLASH MATERIALS

Backsplash options are as varied as those for countertops if not more so, because tile—with its multitude of designs—comes into the mix.

Porcelain and Ceramic Tile

Available in a dazzling array of colors, shapes, sizes, materials, and styles, tile backsplashes are a popular choice. And thanks to advances in printing technology, both ceramic and porcelain tiles can be produced to resemble other materials—such as wood and stone—without the upkeep of the real thing. They're also resistant to scratches, heat, and water.

Marble

While it's hard to beat the natural beauty of marble, the material is porous, so it needs sealing and periodic resealing to prevent staining. Plus, it scratches more easily than other backsplash materials. There are, however, many marble slabs with streaks and patterns that can help hide any imperfections.

Granite

Polished granite is popular for traditional kitchens, while honed granite has a matte finish that's often a popular choice for contemporary spaces. Granite is easy to clean, hard wearing, and available in a range of different colors. It is among the costlier backsplash options, and because it's porous, granite needs to be sealed.

Engineered Stone

Also referred to as quartz, engineered stone is made of crushed quartz mixed with resin. This material is durable, scratch resistant, and nonporous, which means it won't stain. And it's easy to clean with warm, soapy water and comes in a wide range of colors and patterns, some mimicking natural stone.

Solid Surface

Solid-surface materials are not only available in a diverse color range, but they also offer design flexibility. They can be molded seamlessly into angles

above • Using all-white field tiles from countertop to ceiling may, at first, seem to be an uninspired choice, but using a dark grout can make a dramatic difference. And a dark grout hides dirt better than a light one that matches the tiles.

right • A metallic mosaic backsplash adds a contemporary touch to this kitchen. At the same time, light bouncing off the surface brightens the space.

and curves, so there's no seam between the counter-top and backsplash. In addition, they're nonporous, durable, and easy to clean. In fact, their downsides are few. While it's possible to scratch solid surfaces, those scratches can often be sanded out. The material isn't heat resistant so it shouldn't be installed behind a gas cooktop.

Laminate

If you like the look of stone or wood, but your budget won't accommodate it, consider laminate. An afford-able choice, laminate comes in various colors and finishes—including those designed to look like real wood or stone. Easy to keep clean and water resistant, laminate is a hard-wearing choice. Because it's not heat resistant, though, laminate shouldn't be used behind a gas cooktop.

Glass

For a sleek, streamlined look, glass is a popular choice; it can be installed in large, seamless panels. Be sure to use tempered glass, though. It's harder than ordinary glass and far less likely to scratch. Strong and durable, glass is also easy to clean and install, with the panels either screwed or glued to the wall. As an alternative, glass tiles are also readily available.

Stainless Steel

A stainless-steel backsplash gives a kitchen an indus-trial look. It's not only affordable but is also known for its heat-resistant and hard-wearing attributes. Although it's easy to clean, the material can be diffi-cult to keep that way, as it is prone to scratches, dents, and fingerprints.

Beadboard

Beadboard is an affordable option with a vintage look, perfectly suited for a country-style kitchen. Be aware, however, that cleaning the grooves can be time-consuming. And, it's best to team beadboard with a noncombustible material in areas devoted to cooking.

above · Like their ceramic and porcelain counterparts, shimmering glass tiles are available in a wide variety of styles, from rectangular subway tiles like the ones in this kitchen to mosaic and penny styles.

left · Granite is available in a wide variety of colors, but counters and backsplashes in this kitchen are decked out in basic black—an eye-catching contrast to white cabinetry.

DESIGN OPTIONS

A backsplash is an ideal spot to show off your creative side. No matter what material you choose, there's a wealth of design opportunities at your fingertips.

Even if you opt to match your backsplash to your countertop material, there are design decisions to be made. Will it be a conventional 5-in. to 6-in. backsplash or will you run the material all the way to the ceiling? (If you're replacing an existing short backsplash, make the new backsplash 1 in. taller than the original to hide any drywall damage resulting from the tearout.) But a backsplash need not be squared off at the top; nearly every material can take any shape or form.

If you choose tile, though, the options are all but endless, as it comes in every imaginable shape, size, color, pattern, and texture. Standard field pieces are typically more affordable than decorative pieces, and you can create a great deal of drama by positioning them in innovative ways. (See "Designing with Subway Tiles" on pp. 148–149.) But there's no denying that decorative tiles can play the role of a kitchen's focal-point feature.

above • The way in which backsplash tiles are arranged can make a dramatic design statement. But you can take it to yet another level by using a variety of colors, too.

right • Beadboard creates a classic country backsplash that's both affordable and easy to install. Be sure to check local codes, however, before installing it near a cooktop.

above • The backsplash behind the range in this kitchen is a focal point as much for its shape as for the marble material. Taking a design cue from the motif on the hood, the curvilinear design draws the eye toward it.

left • Behind the sink wall in this contemporary kitchen, a back-painted glass backsplash adds an element of shine in a space primarily appointed with matte finishes. The glass is a practical choice, as well, given the inevitable splashes from the sink.

Putting a Backsplash to Work

A modular rail system in this contemporary kitchen emphasizes the linear shape of the backsplash's tiles.

A backsplash is, more often than not, both fashionable and functional, providing an eye-pleasing element while protecting walls from a cooktop's heat and a sink's splatters. But with a little planning, you can make your backsplash work even harder. If you have a small kitchen, for instance—one that's short on storage and/or counter space—consider shallow wall shelves supported by brackets. Likewise, you might incorporate a tiled-in niche behind the cooktop that can hold cooking supplies or frame a pot filler.

Another option is a modular rail system, which consists of a bar attached to the backsplash. All kinds of components can be hung from the rail, including paper-towel holders, utensil bins, and magnetized knife blocks.

above • In the midst of rough-hewn beams, smooth countertops, and shiny stainless-steel appliances, three-dimensional backsplash tiles do their part to add textural interest to this kitchen, too.

right • The natural veining of marble tiles is even more apparent in this backsplash, where the offset-placed tiles create visual interest and movement.

left · Even the smallest design detail can make a big difference. In this kitchen, pale gray glass tile reaches to the ceiling, creating a quietly sophisticated backsplash. Behind the range, though, a darker shade is used, defining the cooking zone.

below · The wood backsplash that flanks this kitchen's sink appears to run right up behind the wall cabinets. Smartly, though, the homeowner used the waterproof countertop material behind the sink itself.

Designing with Subway Tiles

Subway tiles are typically laid in a horizontal, offset pattern. But there are so many ways this basic tile can be applied; here are just a few of the options.

Offset

Stack Bond

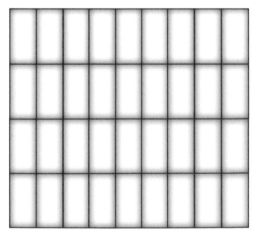

Vertical Stack Bond

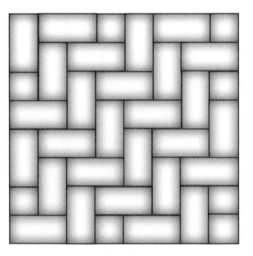

Straight Herringbone

right • Contrasting the pale wood cabinets in this kitchen, a bright blue subway tile backsplash is the definitive focal point.

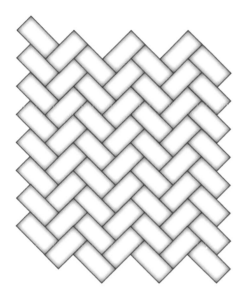

Traditional Herringbone

left • Subway tiles are one of the most unpretentious— and inexpensive—tiles you'll find. But depending on your design direction, the resulting backsplash can be anything but basic.

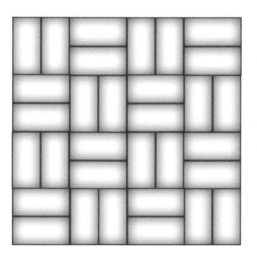

Crosshatch

SINKS AND

Because a lot of time is spent standing at a sink, both it and the faucet

need to be comfortable for your height and your work habits.

FAUCETS

Sinks

Although functionality is their primary purpose, today's sinks have style going for them, too. And if you're not limited by existing cabinetry or counters, the choices are more wide-ranging than ever before.

Start by considering how many bowls you will need, determined by your daily tasks. Will a main sink be sufficient or would you prefer to have a prep sink, as well? Even if a single main sink is your preference, does it need to consist of one, two, or three bowls? If you regularly wash large items in your main sink, for instance, a single or large-small bowl combination (often referred to as a one and three-quarters) might be ideal.

Once you've decided how many sinks you want, consider the material. Stainless steel, for instance, has a contemporary look but can scratch and show water marks. On the other hand, cast-iron sinks are classic but may require a little elbow grease to stay bright white.

Finally, give some thought to your sink's installation. Undermount sinks are attached to the underside of a countertop, creating a clean look, whereas drop-in sinks are installed on top of the counter. Apron-front sinks—also referred to as farmhouse sinks—are typically wide and deep, identified by their broad front edge. And integral sinks are made from the same material as the counter itself, usually fabricated as a seamless unit with the countertop. The style of your kitchen will give you a clue as to which installation is right for you.

above • Perfectly appointed for the gourmet chef, the main sink of this kitchen has a standard faucet as well as a tall gooseneck model well suited for pots and pans. Meanwhile, a prep sink on the island is conveniently close to the industrial-style range.

right • This undermount sink allows the marble countertop and backsplash—as well as the sparkling chrome faucets—to take the spotlight.

above · There's perhaps not a more classic sink to be found than a white farmhouse style, right at home with almost any cabinet, counter, or backsplash.

left · Sometimes a simple solution is best, epitomized by this heat- and stain-resistant stainless-steel sink with a single-handle faucet with a pull-down sprayer.

THE MAIN SINK

Traditionally, a main sink has been located in front of a window where there is a good view. Many homeowners still prefer that placement but, if you do, be sure to size your sink to the window above it. Sink and window widths don't need to match exactly, but the two elements should be similar in width; visually, it can look unbalanced to have a wide sink beneath a skinny window or vice versa.

Another option is to position the main sink in an island; a sink that overlooks the kitchen or family room can make it easier for the chef to converse with friends and family. Whatever placement you choose, however, be sure to allow 2 ft. on each side of the sink, one side for dirty dishes and the other for clean.

The stainless-steel sink and faucets are intentionally low-profile in this kitchen, making it easy to focus your attention on the stunning outdoor view.

Sink Materials

Today's sinks are available in a vast variety of materials, these being the most popular:

- **Stainless steel** is undoubtedly the most popular sink material, with heat and stain resistance going for it. It's also available in a wide range of sizes, configurations, and prices. Look for a brushed or satin finish because polished stainless steel is harder to keep clean.

- **Fireclay** sinks, as their name implies, are manufactured from clay fired at an extremely high temperature. They are highly resistant to scratches, staining, and chipping, and cleanup is a snap; all you need is soap and water or, at the most, a mild abrasive cleanser.

- **Copper, nickel, and bronze** sinks can be stunning, but the cost is high, sometimes prohibitively. Plus, their upkeep is far more demanding than for stainless steel. Copper will oxidize to a dramatic degree, requiring regular polishing or delivering the beauty of an ever-changing patina.

- **Solid-surface** sinks can be integrated into countertops for a smooth, streamlined appearance that's easy to clean. Although they aren't cheap, they have their advantages. Solid-surface sinks are nonporous and stain resistant; plus, small scratches won't show up (because the material is homogenous), and modest scratches can be sanded out.

- **Cast iron** is clad in a tough enamel finish, easy to clean and difficult to scratch. Keep in mind, however, that cast-iron sinks can be unforgiving on dropped glasses. This material is heavy, too, so your cabinetry needs to be structurally sound.

- **Stone** can be carved into the shape of a sink, but a stone sink is more likely to be fabricated from flat slabs, creating an apron-front configuration. Soapstone, for instance, is a traditional sink material that's easy to work with and relatively easy on dropped dishes. Remember, however, that any type of stone that you choose for your sink will require additional support.

- **Engineered stone** sinks comprise a mix of powdered stone (usually quartz or granite) and a small amount of resin. These sinks are tough, strong, don't need sealing, and won't stain.

- **Ceramic** sinks have a smooth, impervious surface and are easy to clean. But they are heavy, which is hard on dropped dishes, and more susceptible to heat than other sink materials.

This cast-iron, double-basin sink is supported to a large degree by its self-rim. Because the backsplash is incorporated into the design, the sink is easy to clean, too.

left · If you're looking for a kitchen sink in a dark neutral hue, consider slate or soapstone. You'll also find models made of a strong blend of cement and jute fiber that's considerably lighter than you'd expect.

THE PREP SINK

There are numerous benefits to a second sink. The main one can be dedicated to cleanup, while the other can be designated as a prep sink. As a result, the area around each sink is given a specific purpose. Cutting boards, knives, bowls, and colanders can be kept near the prep sink, while dishes, silverware, and glasses can be stored near the cleanup sink and dishwasher. Having two separate zones like this improves both traffic flow and organization.

The best location for a prep sink is a spot that's close to both the cooktop and your main work surface, with a substantial amount of counter space adjacent to the sink; think in terms of an area that measures 4 ft. wide by 2 ft. deep. The most effective prep sink is a single-bowl style that's 16 in. to 21 in. wide, large enough to fit your colander and your most frequently used pots and pans. (Most prep sinks also function as secondary cleanup areas.)

Undermount sinks are commonly installed in natural- and engineered-stone countertops. One advantage of an undermount is that you can easily sweep food scraps from the counter right into the sink.

The best location for a prep sink is a spot that's close to both the cooktop and your main work surface, creating a short distance to carry heavy, water-filled pots back and forth.

above • A prep sink in a butler's pantry can be a welcome addition, whether the space is utilized as an auxiliary kitchen or as a wet bar.

left • This contemporary prep sink is a work of art in its own right. It also gets high marks for its position at the end of the island, which leaves the rest of the counter to be used as work space.

left • By moving the sink to a corner of the island, you instantly double the workspace around it. This one, dubbed the SocialCorner, has optional accessories; colanders, cutting boards, and drying racks make it even more efficient.

THE TWO-SINK KITCHEN

In a kitchen where two cooks work together on a regular basis, two sinks make perfect sense. With just one sink, prepping and cleanup activities all have to take place in the same spot, and—when those tasks happen simultaneously—the result is a cramped work area. Two well-placed sinks eliminate this dilemma; there's room for two people to work without one getting in the other's way.

By the same token, however, a two-sink setup works well for one. It prevents dirty dishes from stacking up, forcing the chef to clean out the sink before it can be used again. Instead, dirty dishes go straight to the designated cleanup sink—and the chef can keep right on going.

above • Two sinks in this kitchen—each with its own "territory"—have their own purposes. With a gooseneck faucet, the main sink stands ready to wash even the largest pots and pans, while the smaller sink provides a place to rinse food before prepping.

left • The main and prep sink in this kitchen are staggered—not opposite each other—so that two cooks working at the same time don't get in each other's way.

above • The main sink in this kitchen is crafted of the same durable stone as the countertops, but the prep sink is no less hard-wearing. It's located next to the island's chopping block, making it easy to move washed fruits and vegetables to the cutting surface.

left • The main sink in this kitchen takes a traditional position under the window, while a prep sink is placed at the far end of the island, so few steps are involved when carrying large pots of water to the cooktop and back.

Sink Styles

TOP-MOUNT/DROP-IN/SELF-RIMMING

$

- Weight is supported by a rim that extends above the countertop surface
- Available in stainless steel, porcelain, and enameled cast iron
- Most commonly used with plastic laminate and wood countertops, as they have a vulnerable edge
- Widely available in home centers
- Easy to install, even for do-it-yourselfers
- Pre-cut holes on the sink edge accommodate faucet
- Top-mount stainless-steel sinks are typically made of thinner materials, thus are noisier—and harder to clean—than their undermount counterparts

UNDERMOUNT

$$ to $$$

- Installed below the countertop, offering a seamless look
- Typically used with stone or quartz countertops
- Available in home centers and plumbing specialty stores
- Broadest selection of all sink types in terms of materials, styles, configurations, and sizes
- Faucet typically mounts on countertop
- Require professional installation
- Stainless-steel models crafted in 16- and 18-gauge materials are higher quality—and quieter—than drop-in styles
- Many specialty sinks, such as triple-bowl or modular styles, are made for undermount installation

left • A drop-in sink, like this stainless-steel model, is one of the easiest options for the do-it-yourselfer.

below • An integral concrete sink like this allows the option of incorporating a drain board, too.

INTEGRAL

$$ to $$$

- Made from the same material as the countertop they're set into
- Typically paired with solid-surface, quartz, or stone countertops, though concrete and stainless steel are other options
- Seamless style makes for easy cleanup
- Allow for creativity in terms of design
- Integral drain board can be designed into countertop

FARMHOUSE/APRON-FRONT

$$$

- Protrude from the countertop and cabinetry, making it easier to get close to the task at hand
- Typically found in plumbing specialty stores
- Require a 36-in. cabinet, so may not be practical to retrofit into standard sink base cabinet
- Generally deeper than undermount and top-mount sinks
- Single-bowl styles are most common but double-bowl styles are offered, too
- Traditionally made of porcelain, but today's options include a wide variety of materials

above • **By its very nature, an undermount sink is understated, allowing the faucet of your choice to take star status.**

left • **Most farmhouse, or apron-front, sinks are traditional in style but this one—made of stainless steel—has a contemporary feeling.**

The Farmhouse Sink

To make room for bulky pots and pans, more and more homeowners are opting for large, single-basin sinks as opposed to double-basin styles. And among the most popular is the classic apron-front, or farmhouse, sink. The most common material for farmhouse sinks is white porcelain but—as these sinks continue to enjoy a resurgence of popularity (they originated at a time when there was no running water, designed to hold large amounts of water fetched from nearby wells, lakes, or streams)—you'll also find them in fired clay, stainless steel, copper, stone, and even wood, such as teak and bamboo.

1

1. It's not often that a sink becomes a focal point of a kitchen. Here, though, a stunning stone countertop material forms a farmhouse sink, too.

2. It seems only fitting that the façade of the sink in this traditional kitchen should be detailed in much the same way as the cabinetry.

3. In this kitchen, stainless steel gives the classic farmhouse sink a contemporary attitude, emphasized further by a backsplash that features the same metal.

4. A copper sink was chosen for this kitchen, in part because it blends so quietly into the wood countertop that surrounds it.

5. The front of this farmhouse sink takes a decorative tack, inserting the owners' personality in an unexpected place.

Faucets

It's important not to treat faucets as an afterthought; they should be chosen at the same time that you select sinks and countertops, not only to coordinate styles and colors but also—more importantly—to make sure dimensions are compatible.

FEATURES

One of the first decisions you'll need to make is what type of faucet will best suit your daily routine. Are you looking for a pull-out sprayer with multiple flow settings or is a high-arcing restaurant-style—good for cleaning large pots—a better choice? Is a hands-free or tap on/off faucet important to you, one that will reduce the cross-contamination of food? Are you looking for a matching soap dispenser, a filtration system, or both?

TYPES

Today's kitchen faucets vary widely in style. If you prefer a clean sink deck without a sidespray, a *pull-down* model might be right for you; with one fluid, downward movement, this faucet can reach the bottom of a deep sink. *Pull-out* faucets function in the same way as pull-downs, although pull-out models have a larger section that pulls away, making them easier to grip. A *two-handle* faucet, often the choice for traditional kitchens, allows the user to control hot and cold water from separate handles, while a high-arc *gooseneck* faucet—popular for its industrial-style good looks—allows a tall pot to easily be set in the sink.

above • The spout of this pull-down faucet can reach to the depths of the prep sink, making it easier to wash everything from vegetables to cooking essentials such as bowls and colanders.

right • A high-arc gooseneck faucet is preferred by many homeowners for its industrial look, but it's also a smart choice if you regularly wash large pots and pans.

facing page • Two-handle models, which allow you to control hot and cold water separately, are often the faucet of choice in traditional kitchens like this one.

above • Unlike a pull-down faucet, in which only the spout can be extended, a pull-out version like this one has a larger section to grip, making it easier to handle.

MOUNTING

Faucets can be sink-mounted, deck-mounted, or wall-mounted, each option with its own pros and cons. If you opt for a sink-mounted faucet, be sure that it's compatible with your sink: Check how many mounting holes your sink has and that your faucet not only has the same number but also aligns perfectly. There's more latitude with deck-mounted faucets, as they are installed directly to the counter-top instead of the sink. If you are installing an under-mount sink, for instance, a deck-mounted faucet is an option. Just be sure to allow at least one finger's width behind the faucet for cleaning purposes. Finally, it's especially important to be sure that a wall-mounted faucet is compatible with the sink. The distance that the water spout projects will deter-mine whether the two will be compatible or not, and it can be even more of an issue with a double sink.

MATERIALS

Your faucet's finish may be oil-rubbed bronze, stain-less steel, polished chrome, brass, nickel, or a variety of other looks. As a rule, a brushed finish makes for easier cleanup than one that's shiny, but there's something about a gleaming faucet that still makes it a popular choice.

A faucet's interior mechanisms are typically made from solid brass, a brass blend, or stainless steel. Solid brass and stainless steel hold up better than blends to high pressure and hot water but they also cost more. The faucet's moving parts, which control water temperature and flow, are likely to be ceramic disks. The better quality they are, the smoother they glide and the more control you have over volume and temperature.

right • Many of today's modern faucets have sprayers built in, but in this vintage bridge design, the brass hose is visible at all times—which is part of the charm.

facing page • When placing a deck-mounted faucet like this one, it's important to leave enough room between the faucet and the backsplash to make the space easy to clean.

A Perfect Faucet Fit

- For an undermount sink, position faucet holes so they clear the sink's concealed edge.

- For a drop-in sink, match the faucet and accessories to the number of holes drilled in the rim. If you have too many holes, choose a faucet with a base plate that can cover them or add accessories such as a soap dispenser.

- Consider the clearance behind a sink. Do faucet controls need room at the back for lever handles? Does the spout location pose any problem?

- Deck-mounted faucets on extra-thick counters—such as those made of stone, wood, or concrete—require shank lengths that accommodate the thickness, or compatible extenders.

- In general, a faucet needs sufficient reach. Does the spout swivel in a wide enough arc to reach the entire sink? Two-handle faucets, including bridge-styles, might not work with some double-bowl sinks.

above • The oil-rubbed bronze faucet paired with this prep sink was chosen as much for its traditional style as it was for the color, also found in the stone countertop's markings.

left • Although fingerprints and water marks show up easily on chrome, there's something about the gleaming metal that gives the kitchen a clean feeling.

left • If you're short on countertop space, a wall-mounted faucet can be a good alternative to more conventional deck-mounted styles.

The Extras

Adding a few extras to your faucet or sink can make your kitchen even more efficient. Even better, these add-ons won't take much of a bite out of your remodeling budget.

POT FILLERS

These special-purpose faucets are typically mounted near the cooktop, making it easier to fill large, heavy pots at their point of use. That said, if your food requires draining, you'll still have to carry the heavy pot to the sink, so it's best to be sure a prep sink or main sink is just steps away. As an alternative, some pot fillers are positioned on the countertop next to the prep sink. Pot fillers are typically added during full-scale remodels, as the right plumbing needs to be in place.

SINK ACCESSORIES

Many of today's sinks offer the choice of widely ranging accessories, making cooking and cleanup easier than ever before. Depending on what suits your style best, you might opt for something as simple as a bar to hold a dishrag or an insert where you can set a sponge all the way up to cutting boards, bottom grids, rinsing baskets, utensil trays, and colanders.

above • Pot fillers are typically wall mounted over a cooktop, so a large pot of water doesn't have to be carried from the sink.

left • Made to perfectly match your sink's measurements, a rinse basket like this one eliminates the need for a separate colander.

left • If your cooktop is located on an island, a deck-mounted pot filler makes sense. Likewise, a deck-mounted model can be placed next to an island's prep sink.

below • Many of today's sinks have optional accessories. This undermount, double-basin model, for instance, features a bottom rack that keeps the bottom of the sink from scratching and a cutting board that perfectly fits over the smaller bowl.

SPRAYERS

A faucet and sprayer can be installed as separate elements or as a single unit. For the latter, the head of the faucet pulls down to become a sprayer attached to a flexible hose, extending the reach of the faucet significantly. Most faucet sprayers change from stream to spray with the touch of a button.

WATER FILTERS AND HOT-WATER DISPENSERS

If purified water is important to you—or instant hot water, for that matter—consider a water-filtration system or hot-water dispenser at your kitchen sink. Filtered water can be dispensed through your main kitchen faucet or through its own spout. A hot-water dispenser, on the other hand, is generally a separate unit. And it's useful not only for hot coffee and soup; hot water added to dishes with baked-on food makes cleanup easier. Both hot-water and water-filtration dispensers require the installation of equipment in the cabinet below the sink. If space is limited, it's best to consult with a professional.

HANDS-FREE FAUCETS

Hands-free faucets are available in motion-sensor styles or with tap on/tap off technology. Motion-sensor faucets require only that you swipe your hands below the spout; water then flows at whatever temperature setting you used last. With tap on/tap off varieties, the faucet can be touched anywhere on its body with an elbow or arm to start the water flow—beneficial when your hands are coated in flour or when you're carrying a hot pot and don't have a free hand. Another hands-free option is a foot pedal, which can be locked on for a continuous stream, and then unlocked to control the water with your foot again.

left • Touchless faucets are an asset when you have your hands full. This motion-sensor style can be turned on and off with the simple wave of a hand.

below • An instant hot-water dispenser is a convenience for instant coffee, tea, or soup. Some models offer filtered near-boiling and cool drinking water all in the same system.

above • While many homeowners prefer faucets with built-in spray functions, there's much to be said for separate sprayers. When located in the holder, this one angles into the sink, allowing hands-free operation.

left • More and more of today's homeowners are opting for separate water filters, many of which match—or are very similar to—the main faucet.

right • This single-lever faucet has a built-in water filter, eliminating the need for an additional spout.

FLOORING

A kitchen floor needs to be serviceable; it should resist stains and water,

stand up to heavy traffic, and be easy on the feet. That's a lot to ask,

but you'll want your kitchen floor to be stylish, too.

Choose Your Material

When selecting flooring for your kitchen, consider three basic criteria: How does it look, how well does it work, and how does it make you feel—physically—at the end of the day. Is it hard-wearing and easy to keep clean? For durability and longevity, materials such as ceramic and stone tile are long lasting. But on the downside, neither is particularly comfortable to stand on for long periods of time. If comfort is your key criteria, choose a softer floor material such as wood; it's not only easy on the feet but adds warmth to a room, too.

If, during the remodeling process, you'll be changing the kitchen floor material, be sure that it complements the flooring in adjacent spaces. Choose tile for the kitchen that will coordinate with hardwood in a nearby dining room, or wood in the kitchen that will harmonize with the carpet in your living room. Meanwhile, an open floor plan can make things easier; simply run the same floor—like a handsome laminate—throughout the kitchen and great room.

Finally, when making your flooring purchase, order all that you'll need for the project so each piece is from the same production process, plus 5% to 15% more material than your design requires. That way, you'll have extra pieces for future repairs, if needed.

above • Some of today's trendiest kitchens are decked out in gray. In this kitchen, for instance, gray cabinetry and stainless-steel appliances complement the gray-stained wood flooring.

right • Light wood flooring in this kitchen is handsome in its own right but, in contrast to the dark cabinetry, it takes on an even more dramatic flair.

above • Wood teams up beautifully with brick in this kitchen. Not only are the materials well suited for the traditional space, but the classic pattern they create is, too.

left • If you're looking for the warmth of wood, the real thing isn't your only option. Also consider engineered woods and high-quality laminates.

Planning Pointers

Replacing the flooring in your kitchen invariably comes with structural concerns. A key consideration, for instance, should be the thickness of the new material. Perhaps you're replacing vinyl tile with solid wood; with the new material will come extra height. Will your refrigerator still fit under the cabinet above once the new flooring is installed? An extra ½ in.—or even less—could make all the difference as to whether the appliance will fit into its intended space. Likewise, new flooring might lock your dishwasher into place, making it impossible to be serviced or replaced without damaging the new surface. Talk to your installer about running the material under at least the front legs, making sure that your machine can be adjusted to accommodate the new height.

If wood or laminate is your flooring of choice, keep in mind that both require expansion gaps that allow the material to expand and contract as the temperature changes. Your existing baseboards may not cover the gaps; if not, address the issue with a professional before investing in the new material. Some companies will install new baseboards with your flooring. Factor in the time and cost of having the baseboards painted, though, which some flooring companies will not do. And if the new material butts up to your existing cabinetry, you'll need molding to cover expansion gaps there, too.

Finally, be sure that your flooring's required maintenance fits your lifestyle. Wood floors, for example, require constant care; spills need to be wiped up right away to prevent moisture damage, which can be a challenge if you have an active family. In that case, reclaimed wood might be a better bet. Any dings from dropped dishes or scratches from pets' claws will only add to the charm of a reclaimed floor's inherent distress marks.

The primary work area of this kitchen is defined by its tile floor. It's easy to stand on for extended periods of time, much like the cork flooring just beyond the space-dividing wood trim.

The beauty of concrete is that it can be colored in myriad ways or left in its more natural gray state—often the preference in contemporary interiors.

above • Stone floors are a hard-working choice for the kitchen, but if there's no room in your budget, ceramic and porcelain tiles can give you the look for less.

left • Reclaimed wood floors have a rustic charm well suited to traditional or country kitchens. They're a good choice, too, for active families; additional dents and dings will only enhance their appeal.

Woods and Their Look-Alikes

Wood floors—and their look-alikes, for that matter—continue to gain in popularity. Which one is best for you comes down to aesthetics, maintenance, and, of course, costs.

WOOD

Because it's warm and resilient, and can be refinished again and again, wood flooring is a perennial favorite. The majority of solid wood flooring is oak strip (¾ in. thick and 2½ in. wide); wider planks are available but they are more expensive. Other options include hardwoods such as maple, cherry, and hickory. Softwoods like heart pine and fir are handsome as well, but they are more susceptible to dents and dings. That said, some homeowners like the added distressing.

Solid wood flooring is available in both prefinished or unfinished forms. Finishing wood floors in place takes time, emanates odors, and requires staying off them for a few days. Although this option isn't quite as hard-wearing as a factory finish, it provides better overall protection because the joints are sealed along with the strips. Plus, solid wood floors can be refinished multiple times. In terms of maintenance, wood floors require frequent vacuuming and damp mopping, as well as quick attention to spills. Keep in mind, too, that glossy finishes and dark tones will show scratches and stains more easily than satin finishes and medium tones.

Wide-plank wood floors, as opposed to their narrow-strip counterparts, can make it seem—from a visual point of view—that your kitchen is larger than its actual dimensions.

Hand-scraped floors—available on hardwoods or engineered wood—have a worn, Old-World look with the benefit of modern, protective finishes.

Although pine is a soft wood and dents more easily, many homeowners like the authentic look it lends to traditional rooms.

Bamboo Flooring

Although it's technically a grass, bamboo has much in common with wood. It's a fast-growing plant and has varying degrees of hardness. Strand-woven and end-grain bamboos have superior density and durability, but flat-grain and vertical-grain bamboos are less expensive. Like wood flooring, bamboo can be floating, nailed, or glued, and it can be prefinished or finished in place. It's becoming more sustainable all the time, too; although bamboo was once exclusively imported, some is now being grown in the U.S.

Like any other flooring, bamboo has its pros and cons. On the plus side, top-quality bamboo flooring is as durable as conventional hardwood flooring and, depending on the thickness of the planks, it can be refinished like hardwood, too. Be aware, however, that inexpensive bamboo floor is susceptible to scratches and dings.

ENGINEERED WOOD

Engineered wood flooring has become an increasingly popular choice for several reasons. It has the look of solid wood but consists of a thin layer of solid appearance-grade wood laminated to several layers of plywood, making it more dimensionally stable than its solid counterpart. Additionally, engineered floors are typically less expensive than solid wood and are easy to maintain; a simple sweep of a broom or a vacuum cleaner with a soft flooring attachment is all that's needed. Better yet, installation is easy for the do-it-yourselfer.

Engineered wood floors were first developed for use over concrete slabs, but the thicker, ¾-in.-thick versions can be nailed down over a wood subfloor, the same way you would install a plank floor. Meanwhile, some of the newest, thinnest examples use a tongue-and-groove system that locks them in place. These floors, often referred to as floating floors, can be placed over a cork underlayment or directly over an older floor.

There is a downside, however. Because the top layer is so thin, it can be a challenge to refinish engineered flooring. Whereas solid wood can be refinished numerous times, engineered wood is limited to once or twice.

Laminate: The Great Pretender

Laminate is a great pretender; it can simulate the look of wood, tile, stone—almost any hard-surface flooring, for that matter. It comprises a clear wear layer that protects a photo layer. These two layers, in turn, are laminated to a high-density fiberboard and a moisture-resistant melamine bottom layer. Today's laminate flooring planks or tiles float on a smooth underlayment and typically snap together tightly with no glue required. It's a relatively economical flooring, easy to install, comfortable to stand on, and easy to clean, too. For the most realistic-looking flooring, look for surface embossing that matches the texture you see in the photo layer.

Laminate flooring is susceptible to scratches, however, and cannot be refinished. As a result, it's a good idea to buy extra with your initial purchase, in case you need to replace planks or tiles at a later date.

left · Compared to hardwoods, engineered wood and laminate floors can be more resistant to moisture, making them a good choice for the kitchen.

above · Advancements in the industry have made it harder than ever to distinguish between true wood flooring and its look-alikes, engineered wood and laminates.

left · A conventional wood color would have been a more traditional approach in this kitchen, but the dark hue creates a clearly contemporary look.

Tile

Stone, ceramic, and porcelain tiles are classics in the kitchen; they're elegant, durable, and last a long time. They come in a wide variety of shapes and sizes, too, providing near endless design possibilities. That said, not all tiles are created equal; the tiles you choose must be rated for floor use, with enough hardness and slip resistance to serve their intended purpose.

STONE

Stone tiles are created from quarried materials, with granite, marble, limestone, travertine, and slate being particularly popular. Prized for their luxe look, stone tiles are typically installed in large-format sizes with nearly imperceptible grout lines, creating the appearance of a natural stone floor.

If you live in a warm climate, you'll no doubt appreciate the fact that stone is cool to the touch. Conversely, if your winters are cold, radiant-heated stone floors can be a welcome amenity; stone conducts heat well, making it a good choice for radiant heat systems. Stone is expensive, however. Choosing one that's quarried locally is one way to cut down on costs, as shipping increases the price substantially, but even a locally quarried stone floor will cost more than other options, such as wood. Stone flooring requires professional installation, too, as well as a substrate that can handle the additional weight. Finally, be sure to factor in maintenance. All stone tiles, with the exception of soapstone, require regular sealing to resist staining.

above • Stone tile, like this slate, adds a natural quality to a room that can't be found in any other flooring material.

left • The inherent appeal of stone in this kitchen is further enhanced by the assorted sizes—and random placement—of the tiles.

right • Because stone comes directly out of the earth, there can be wide-ranging variations in its color. But that is much of what makes the material so attractive.

The looks you can fashion from stone tile are all but endless. This basketweave pattern, for instance, is the result of combined mosaic rectangles and squares.

Radiant Heat

If you plan to replace your kitchen floor, consider adding radiant floor heating at the same time. Because it heats a room from the bottom up, it's ideal for naturally cool floors like tile.

Two types of radiant floor heating systems are recommended for residential use—*electric*, which uses cables, and *hydronic*, which uses water tubes. An electric system is typically used if you're going to install it in only one room, although it is more energy-intensive. You can reduce costs, though, by programming your system to operate only when you need it, using a thermostat independent of your home's forced-air system. Hydronic systems, on the other hand, rely on heated water that circulates through PEX (cross-linked polyethylene) tubing. Both types are embedded beneath the finished flooring.

Radiant floor heating systems are more expensive than their conventional forced-air counterparts but, over time, save money due to better efficiency. (There aren't any ducts to leak air and waste energy.) Plus, there's no blowing air, meaning less stirred-up dust and fewer allergens.

CERAMIC AND PORCELAIN

Like stone, ceramic and porcelain tiles are created from natural materials. They're manufactured with different types of clay, but the real difference comes in their water absorption. Specifically, porcelain tiles absorb less than 0.5% of water, whereas ceramic tiles absorb more than 0.5%. That means porcelain is a denser material, making it a no-brainer for moisture-prone areas like the kitchen. That's not to suggest that ceramic tile is a bad choice, just that porcelain is slightly better.

Both ceramic and porcelain excel in high-traffic areas, but because porcelain is denser it offers better long-term resistance to scuffs and scratches. Through-body porcelain—in which the color goes all the way through the tile—is especially scratch resistant, a quality busy households will appreciate. And both types of tile come in a wide range of sizes, colors, and patterns; today's screen printing and ink-jet processes can make their surfaces nearly indistinguishable from wood, terracotta, and even stone.

When choosing ceramic or porcelain tile, look for the largest-size tile that works with the scale of your room; fewer grout lines will mean less potential staining, though you will need to seal the grout before the floor is used and periodically afterward. If your budget permits, consider a tile with rectified edges, which allow tiles to be installed so closely together that grout lines can be as thin as $\frac{1}{16}$ in.

There are, of course, drawbacks to a porcelain or ceramic tile floor. First, it's hard on your feet and your back—and on dropped dishes, too. And some tiles, especially polished styles, can have a high slip factor; check with your installer about your material's slip resistance. Finally, costs come into play. Although it has its benefits, porcelain tile is generally more expensive than ceramic. If you're on a tight budget, though, you're sure to find a ceramic style to fill the bill.

By pairing solid field tiles with patterned styles, you can establish the look of a soft rug on your hard-surface floor.

above • Floors with complex motifs don't have to be difficult to install. Patterned porcelain and ceramic tiles can be placed side by side to create intricate designs like this one.

right • For those who like wood but want something easier to maintain, plank-size tiles that replicate the real thing can be a good alternative.

above • Today's tiles are available in an amazing array of colors and patterns, shapes and sizes. Terra-cotta tiles, for instance, could be a good fit for this Old World-style kitchen, but so could ceramics that replicate the look.

Concrete

If durability is a top priority, concrete might be your flooring of choice. It's also sustainable and budget-friendly, works well with radiant heat, and comes in a wide array of colors and textures.

The look of concrete is determined by the color and the aggregate particle mix that makes up the actual concrete. Color can be achieved via the aggregate itself, mixing colored powder or liquid into the aggregate, acid etching, staining, or using a tinted wax. Both the color and texture can also be altered by polishing—which exposes more of the aggregate—and sealing. And adding a coat of wax can create either a gloss or matte finish; just be aware that this process might darken the color. (Sealing or waxing is recommended in high-traffic areas like the kitchen to maintain a protective layer.) Concrete floors are easy to care for, too; all that's needed is a regular mopping with soapy water or a neutral cleaner.

Although the flooring itself is concrete, its rich brown tone warms up this otherwise white—and potentially cold—kitchen. Color can be added to concrete when it's initially poured or later in the form of water-based dyes or acids.

Sections of concrete team up with wood planks to form a one-of-a-kind floor in this kitchen. The resulting design is right in line with the rest of the room's rectilinear look.

Resilients

Resilient flooring is a favorite in kitchens; it's comfortable to stand on, easy to install, and relatively inexpensive. Plus, it can last for years if well maintained, damp mopping to remove grit and wiping up spills from seams. Most resilient flooring comes in both tiles and sheets, while some is available as floating-floor planks or tiles that snap together without an adhesive.

Cork flooring is considered sustainable because the bark is peeled from a live tree without damaging the tree itself. It's quiet and resilient and repels mold and mildew, though it does need to be sealed upon installation and every few years afterward.

Vinyl flooring isn't the shiny no-wax flooring that it once was. Today, it's typically made of a tough outer coating, a clear vinyl layer, a printed design or color layer, and a bottom layer of felt or fiberglass. Vinyl is available in large sheets or tiles; sheets offer more water resistance because there are fewer seams.

Linoleum flooring is also sustainable, hypo-allergenic, quiet, and comfortable to stand on. Today, it's enjoying a resurgence of popularity thanks to its green credentials and old-fashioned charm. Linoleum can be installed in sheets or tiles that can be easily replaced if damaged.

left • Concrete floors have become increasingly popular in kitchens and throughout the house, especially in contemporary interiors. They're durable, easy to maintain, and available in a wide range of textures, colors, and polishes.

Flooring

WOOD AND BAMBOO

$$ to $$$

- Solid wood floors offer long life; can be refinished multiple times

- Solid wood floors are thick; make sure they will fit under your cabinets and appliances

- Available in sustainable species from protected forests; ability to refinish them makes them even more sustainable

- Prefinished, engineered types are available, with protective layer on top, finished wood below, and an unfinished substrate

- Engineered wood can be floated over concrete slab floors

- Oak, cherry, maple, beech, walnut, and birch are offered in solid and engineered-wood varieties

- Wide-plank styles are popular, with 5-in. widths most common; wider varieties—from 8 in. to 12 in.—are available at premium pricing, most typically in engineered wood

- Durable bamboo often looks like wood but is made from fast-growing grasses

- Like wood, bamboo is comfortable underfoot

LAMINATE

$ to $$

- Can take on the look of any flooring material because it consists of a photographic image beneath a protective finish

- Can typically be installed where engineered wood floors work, but is generally less expensive

- Protective finish offers excellent durability, though laminate can't be refinished if damaged

- Typically easy to maintain

- Most laminates are DIY-friendly due to click-lock installation

STONE

$$$

- Popular stone floors include granite, slate, marble, and travertine

- Should be professionally installed for optimum results

- Naturally occurring variations from tile to tile are a feature, not a flaw

- Available in large-format (typically 18 in. or 24 in. square, and 12 in. by 24 in. or 36 in.) and mosaic sheets

- Stone should be sealed for better stain protection; frequency depends on type of stone and traffic level of your kitchen

- Polished natural stone needs to be polished periodically

- Like porcelain and ceramic, stone is hard underfoot

above · Oak floors, or any hardwood with an oak stain for that matter, bring a certain warmth to a kitchen.

right · Stone tiles with an aged look—due to natural pitting and pores—are particularly well suited for traditional kitchens.

Whether your preference is stone, porcelain, or ceramic tile, you'll find that the patterns you can create are all but endless.

As more and more of today's homes are incorporating great rooms, solid wood floors are finding their way into the kitchen. Their handsome good looks can come in the form of engineered woods or laminates, as well.

CERAMIC AND PORCELAIN TILE

$ to $$$

- Ceramic is widely available and affordable
- Properly installed, glazed ceramic and porcelain tile are impervious to surface stains and moisture
- Both are offered in a wide range of shapes and sizes, colors and patterns
- Through-body finishes, most often seen in higher-price-point porcelains, make chips and dings less visible
- Rectified tiles can be placed closer together, resulting in narrower grout lines
- Both are extremely hard to stand on for long periods of time
- Neither glazed porcelain nor ceramic tile needs to be sealed, though grout between the tiles should be sealed periodically for greater stain protection

CONCRETE

$$

- Offered in a variety of colors, patterns, and finishes
- Typically sealed to prevent staining
- Impervious to moisture, making it a good fit for a kitchen
- Hard to stand on for extended periods of time
- Cold to the touch, but can be warmed up with radiant floor heat

CORK

$$ to $$$

- Natural and sustainable
- Available in assorted finishes and styles, from squares to herringbone
- Naturally insect repellent, sound and temperature insulated, and fire resistant; ideal for those with mold, dust mite, and other household allergies
- Comfortable to stand on for long periods of time
- Requires regular cleaning with special products, as dust, dirt, and harsh household cleaners can damage surface
- Needs to be sealed upon installation and every few years thereafter

VINYL AND LINOLEUM

$ to $$

- Widely available
- Easy to install in tile format
- Vinyl tile and sheets are affordable options for budget remodels
- Sheet vinyl and linoleum should be professionally installed for best results
- Both are comfortable to stand on for extended periods of time
- Often confused with vinyl, linoleum is both durable and sustainable, made from linseed oil and other natural ingredients

FINISHING

Whether they're functional, fashionable, or a little of both,

the finishing touches in your kitchen represent a good share of space. As such,

they deserve some careful planning, not to be treated as mere afterthoughts.

TOUCHES

Putting It All Together

When planning a kitchen, it's easy to get caught up in the allure of handsome cabinetry and high-tech appliances. But window, wall, and ceiling treatments deserve some attention, too, as do personal touches—from treasured collectibles to bar stools that reflect your sense of style. Last but not least, the right lighting is imperative, whether it's to cast a generic glow or spotlight a particular piece.

Some of these finishing touches require upfront planning while others can be put off until the project's end. If, for instance, you have a special ceiling treatment in mind—perhaps a coffered style—you'll need to make it part of your overall remodeling scheme. Likewise, plan ahead if you have your heart set on wrapping the kitchen walls with tile. And one of the most important aspects to consider early on is the lighting plan; you'll need to work with a lighting technician to make sure that your kitchen is covered in ambient, task, and accent categories. On the other hand, some selections can be postponed until later. You might benefit, for example, by picking your walls' paint color after the cabinets and floors are in place.

A textured tile wall is an attention-getter in this kitchen otherwise decked out in smooth surfaces, but the wall covering's inherent sparkle makes it even more dazzling.

above · Country touches stand out prominently against this predominantly white kitchen. Windsor-style bar stools pull up to the island, while a pair of barrister bookcases are transformed into convenient storage for everyday dinnerware.

left · In this transitional kitchen, lantern-style fixtures get a modern makeover; their frames are crafted of polished chrome in lieu of more expected brass. Still, this kitchen isn't void of tradition, proven by the wall clock above the cooktop.

Contemporary pendant lights are a good fit for the streamlined style of this kitchen. Their splayed shapes create overlapping pools of light, too, providing enough light for the work surface below.

Wall Treatments

The walls surrounding your kitchen should be aesthetically pleasing and stand up to food and water stains. And surfaces behind the stove need to be heat resistant. The good news is that there is a wide variety of kitchen-appropriate materials. Select colors and textures that will complement your cabinetry and appliances, and—in a space often dominated by one-color elements— perhaps add a touch of pattern.

Tile, stone, and brick are particularly well suited to walls prone to splashes and splatters, as are wallpaper and glossy or semigloss paints. One of paint's best attributes is that it's affordable. If you tire of a certain shade over time, you can paint right over it. And wallpaper is finding its way into more and more kitchens, as long as it's washable.

There's no doubt stone and brick can bring a natural aesthetic to a kitchen. Still, ceramic tile tops the list in terms of popularity. It's easy to clean and can stand up to heat and high moisture. And a wide variety of colors and patterns makes it versatile from a design standpoint. With the ability to create a complex mural or showcase solid colors in a creative way, ceramic tile offers some of today's most stylish solutions.

above • To keep costs down, use tile in select spaces, complementing it with less expensive paint. If you don't find the shade you're looking for, go the custom route; your paint dealer can match any hue.

right • Wood paneling applied vertically is a time-tested wall covering. Here, though, the concept gets a new twist; paneling is applied horizontally and given a bright white coat of paint.

above • Shades of gray are appropriate in this kitchen's wall covering, used throughout the room—except around the range where heat resistance is required. In a space often lacking pattern, wallpaper can be welcome; just be sure that it is washable.

left • White cabinetry, countertops— even the ceiling—make this geometric-patterned wallpaper stand out more prominently. White-framed windows would have been the expected tack, too, but a soft shade of yellow brings yet more attention to the walls.

THE FIFTH WALL

The fifth wall in a room—the ceiling—
deserves just as much attention as the four
walls that rise to meet it or the floor that it
mirrors in size. And given a ceiling's generous dimensions, there's any number of ways
to treat it.

While many kitchen ceilings are painted
white, they can be a canvas for creativity.
Color, for instance, can do wonders. Painting the ceiling a shade darker than the
walls will visually lower the surface, making
the room feel more intimate in the process. Conversely, painting it a lighter color
than the walls will seemingly add height
to a space. But there are multiple options
for a ceiling beyond basic paint. Consider
rustic beams for a country-style kitchen,
beadboard paneling for a cottage feeling,
or wallpaper to add a splash of pattern. You
might even create architectural interest
with a coffered or barrel ceiling.

above · Rustic ceiling beams give this
kitchen a country ambiance, their
substantial size visually lowering the
ceiling. That, coupled with the room's
fireplace, results in a cozy feeling.

right · This clean-lined, white kitchen
could have gone in any design
direction, but a coffered ceiling
teamed with beadboard paneling gives
it a cottage-style quality.

Not only does this kitchen have a cove ceiling, it's coffered, too. The resulting sections provide a place for evenly spaced recessed lighting.

Refined ceiling beams emphasize the soft curve of this kitchen's cove ceiling. Cable lighting is a good choice here; it doesn't interrupt the tiled ceiling's visual impact.

Window Treatments

A kitchen's window treatments do much more than provide decorative touches. These unsung heroes control light, provide privacy, and can even conserve energy. Window treatments in the kitchen do, however, require special consideration. Because they'll be exposed to grease, food, and moisture, any materials you choose should be easy to clean.

Typically made of fabric, Roman shades are a good choice; even when fully extended they hang completely flat and out of the way. In the raised position, though, they create horizontal pleats that look more like a valance, allowing sunshine to flood your space. You'll also find Roman shades made of natural materials, such as bamboo, cane, grass, or various other reeds. Even in their fully extended position, this type of window treatment allows light to softly filter through.

Another good choice for the kitchen is wooden blinds. Although warm brown tones are still perennially popular, you're just as likely to find wood blinds in a variety of colorful paints and stains, accented with plain or patterned cotton tapes. And they can be sized for standard windows as well as those that are arched or angled. The beauty of wooden blinds goes beyond their aesthetic value. The moveable slats—which can range in width from 1 in. to 2½ in. —can be opened to allow maximum light into the room or closed to provide privacy and keep in the heat on a chilly day.

The bottom line is that your choices are wide-ranging. From café curtains to valances, shutters to mini-blinds, there's a window treatment to suit every kitchen's design style.

right • Roman shades aren't only made of fabric; they're also available in natural materials, such as bamboo, cane, grass, or various other reeds.

below • The beauty of a Roman shade is that it can be fully extended to cover a window, completely raised to resemble a valance, or positioned anywhere between.

left · Wooden blinds control light and privacy at this kitchen's trio of windows. The arched transoms above, though, have been left plain, allowing a better view of their striking silhouettes.

left · In a kitchen, typically dominated by hard edges, curtains can lend a soft touch. Just be sure to keep fabric window treatments away from heat sources.

Lighting

A well-lit kitchen does more than simply shed light on your everyday tasks; it can—and should—be well balanced, using a mix of ambient, task, and accent lighting. Here's what to look for in each type:

Ambient lighting provides overall illumination, making it easy to move throughout a room. In the daytime, natural light can provide much of the ambient light a kitchen needs; at night, it's provided by pendants or chandeliers, recessed cans, and cable or track lights.

Task lighting illuminates a work area, whether it's a countertop, cooktop, or sink. Undercabinet fixtures can provide task lighting, as can pendants, track lights, and recessed can lights. Keep in mind, however, that task lights positioned close to their respective workspaces are more efficient than those positioned on the ceiling.

Accent lighting focuses on specific objects, such as artwork and collectibles, or it can wash a wall with light. Sconces and track lighting, as well as in-cabinet fixtures, can all provide accent lighting.

Even the best-laid lighting plan, however, will benefit from some flexibility. It's a good idea to use dimmers, switching various sources of light separately to suit different moods—and to save energy, too.

left · Recessed fixtures in this kitchen's vaulted ceiling lend ambient lighting; they're too far away from any work surface to illuminate specific tasks. Pendants, on the other hand, are close enough to the island to shed sufficient light on the job at hand.

facing page · Suspended from the ceiling, a contemporary fixture provides both task and ambient lighting in this kitchen. Meanwhile, wall-mounted downlights over the sink deliver task light in that work zone.

below · The primary work zones in this kitchen are around the perimeter, so it makes sense that task lighting—in the form of recessed fixtures—is positioned around the perimeter, too.

above · Wall-mounted downlights flank an oversize window in this kitchen, washing the walls below with light and, in the process, bringing more attention to the impressive window.

RECESSED AND TRACK LIGHTING

Because they can be positioned where you need light most, two types of fixtures are particularly good choices for a kitchen: recessed and track lighting.

Recessed downlights have been a kitchen standard for decades, partly because they come in a wide variety of shapes and sizes, with just as many bulb, baffle, and color options. By spacing downlights close enough together, light pools will overlap on your counter or floor. Keep in mind, though, ceiling-mounted downlights will be relatively far from a workspace, so their bulbs need to be bright enough to produce enough light for safe working conditions. Adjustable recessed downlights, on the other hand, make ideal accent lights for art and decorative collectibles.

Track lighting comes in all shapes and sizes, including high- and low-voltage systems with pendants and adjustable spots. Track lights can be individually adjusted and offer the full range of lighting functions, from ambient to accent. Cable systems are similar to their track counterparts in that individual fixtures can be positioned where they're needed, but these fixtures run on thin, suspended cables rather than flat tracks. Monorail lighting systems are also similar to track lighting but can be bent into curves, giving you more design flexibility.

A cable track system stretches across the vaulted ceiling of this contemporary kitchen. The beauty of this option is that individual fixtures can be positioned where they're needed most.

above · It makes sense to use recessed lights over a work area like this sink; they shine light downward, directly onto the task at hand.

above · A cable system provides ambient lighting in this kitchen, while recessed under-cabinet lighting over the window makes the counter below a better place to work.

right · Recessed fixtures positioned at regular intervals give this kitchen an evenly balanced, overall glow. The ambient light is supplemented by a trio of pendant lights over the island, which double as task lights for the work surface.

LED Lighting

Not so long ago, LED (light-emitting diode) bulbs and fixtures were a new concept. But they've quickly become the standard, representing a leap that seems light-years beyond the conventional light bulb. In the kitchen, LED technology can serve in several ways:

- Standard bulb configurations can replace halogens in recessed ceiling lights or incandescents in pendants and chandeliers.

- Task-lighting fixtures can illuminate cabinet interiors or attach under cabinets to shed light on countertops.

- LED bulbs are compatible with decorative lighting fixtures, including wall sconces and pendants.

In addition to the myriad ways it can be used, LED technology is far more energy efficient than its incandescent and compact fluorescent counterparts. Plus, LED bulbs can last for 10 to 20 years—a definite advantage for ambient room lighting in kitchens with high ceilings.

Pendants and Chandeliers

Pendants and chandeliers can provide ambient or task lighting in a kitchen, and an element of style at the same time. Both are good choices to illuminate a dining table, and two or more can sufficiently light an island. Choose fixtures that are easy to clean, especially if they're located in the kitchen's work zone and height is an important consideration. As a rule, hang a pendant or chandelier 30 in. above a dining table and 36 in. above a countertop. To allow more ambient light to reach the ceiling, choose a semi-opaque, translucent, or transparent shade; if it's task lighting that's needed, select a fixture with an opaque shade that will direct the light downward.

1. In this elegant, tone-on-tone kitchen, light fixtures are equally stylish. A pair of cone-shaped pendants illuminate the island, while a chandelier—not a match, but equally understated—sheds light on the nearby table.

2. Many kitchens are all-white, allowing elements like bar stools and rugs to introduce color. This one takes the concept to new heights, however; a pair of bright yellow pendants is the indisputable focal point.

3. Twin islands in this kitchen inspired pendants that are identical, too. The finish of the polished chrome fixtures takes its cue from the bar stools' nailheads and the cooktop hood's trim, resulting in a fresh twist on traditional style.

4. A brass pendant over the sink and a beaded chandelier over the island are completely different in style. Still, they complement one another because they both have gently curved silhouettes.

SKYLIGHTS

A kitchen's natural light isn't limited to windows and doors. Look up: Skylights can effectively flood a room with sunshine. In some ways, in fact, they have an advantage over conventional windows. They allow an overhead view of the sky and trees while providing privacy at the same time.

Another benefit of skylights is that they can reduce energy bills. While too many skylights can have the opposite effect—letting in too much sun during the summer months and allowing heat to escape in the winter—the right balance of windows and skylights can be a great advantage.

Although skylights can be added to existing spaces, installing them during new construction—or a complete remodel—is, for the most part, preferable. If you do plan to add skylights to your existing space, be sure to get the best you can afford; it will be worth the cost in the long run. Lower-quality or improperly installed skylights can leak, causing headaches in the future.

above • Skylights set into the vaulted ceiling of this kitchen direct light onto the island in the center of the space. They're supplemented by pendants with glass globes that practically disappear against the white ceiling.

left • A series of windows in this ceiling's peak allows sunlight from all directions. Once the sun goes down, though, six galvanized fixtures provide light from above.

right • The central portion of this ceiling is defined by wood paneling, interrupted by a square, oversize pane of glass that—in addition to daylight—allows a view of the roof's peak.

below • A coffered ceiling spurred the idea of this skylight; a center panel of the ceiling was replaced with a window. It provides plenty of natural light during the day, but at night pendants on either side of the skylight take over.

Personal Style

You've chosen your cabinets and appliances, countertops and flooring, even window and wall coverings. But a kitchen isn't complete until it's been given some of your own personal style. There are countless ways to accomplish this, both large and small. The only prerequisite is that the elements you choose are selected more from your heart than your head.

Some personal touches will need to be thought through at the beginning of your design process. A tile mural behind your cooktop, for instance, will require pre-planning. On the other hand, cabinet hardware that reflects your personal interests—like the beach or Mother Nature—can be added at the end.

Finally, displaying artwork or collections is a great way to add a personal touch to kitchens, especially those that flow seamlessly into adjacent great rooms. Just be sure that the items you choose are suitable for a kitchen's sometimes steamy, food-splattered environment.

left · A backsplash with a built-in shelf provides a place to set framed pictures. Meanwhile, antique crocks on the wall shelf above speak to one of the homeowner's favorite collectibles.

below · A mural positioned over this sink adds a personal touch and is a welcome relief from the kitchen's otherwise hard edges.

above · The slightly distressed surfaces in this kitchen could have taken it in a contemporary direction, but vintage accents give it a traditional turn.

left · A wall-hung chalkboard serves as this family's command center. Meanwhile, the shallow countertop below is a good place for after-school snacks or doing homework.

Collectibles

As much as any other accent, collectibles can speak to your personal style. When showcasing them, don't break up the set; pieces grouped together make a more dramatic impact than if they are scattered throughout the room. The collectibles themselves may offer clues as to how to present them, too. Items that are graduated in size or color, for instance, have a built-in order.

1. Kitchen collectibles can be decorative, functional, or a little of both. This assortment of cutting boards, in various shapes and sizes, beautifully fits both categories.

2. A grouping of pans, conveniently hung over the cooktop, looks more like collected works of art than the practical pieces they are.

3. These green vintage bowls make more of an impact grouped together than if they'd been placed throughout the kitchen. The red-and-white graphic backdrop also does its part to increase the drama.

4. A treasured collection of blue-and-white pitchers is clearly visible at the end of this island but out of harm's way at the same time.

CREDITS

p. i: Hulya Kolabas, design: Raquel Garcia Design

p. ii: Jim Westphalen Photography/ Collinstock, design: Lillian August Design (architect: TruexCullins Architecture; builder: Roundtree Construction)

pp. iv–v (left to right): Helen Norman; Matthew Quinn, Home Refinements; Rob Karosis Photography/Collinstock, design: PKsurroundings; Tria Giovan

p. v: (left to right): Helen Norman, architect: Ken Pursley, styled by Rebecca Omweg; Andrea Rugg Photography/Collinstock, design: Anchor Builders; Helen Norman, design: Lauren Liess; Helen Norman

p. vii: Chipper Hatter, design: Roomscapes

p. viii: Hulya Kolabas, design: Tiffany Eastman Interiors & Susan Glick Interiors

p. 2 (top, left to right): Olson Photographic, design: Callaway Wyeth Architects; Helen Norman, design: Lauren Liess; Trent Bell Photography, design: Elliott & Elliott Architecture; Hulya Kolabas, design: Mar Silver Interiors; (bottom): Ryann Ford Photography

p. 3: Susan Teare, design: Conner & Buck Design Build

CHAPTER 1

p. 4: Mark Lohman, design: Jeff Troyer Architect

p. 6: (top) Tria Giovan, design: Kevin Spearman; (bottom) Emily Followill Photography/Collinstock, design: Frank G. Neely Design Associates; Amy Morris Interiors

p. 7: (top) © Brian Vanden Brink; Group 3 Architects; (bottom left, bottom right) Chipper Hatter, design: Kristianne Watts, KW Designs

p. 8: Trent Bell Photography

p. 9: (top, bottom) Mark Lohman; (middle) Tria Giovan, design: John Bjørnen, Bjørnen Design

pp. 10–11: Hulya Kolabas, design: Raquel Garcia Design

p. 12: Tria Giovan

p. 13: (top) Ryann Ford Photography, design: Emily Seiders, Studio Seiders; (bottom), Trent Bell Photography for Eric Allyn Architecture

p. 14: Tria Giovan, design: Kevin Spearman

p. 15: (top) Tria Giovan, design: Courtney Hill; (bottom) Hulya Kolabas, design: Mar Silver Interiors

p. 16: Andrea Rugg Photography/ Collinstock, design: Chester Hoffman Associates

p. 17: (top) davidduncanlivingston. com; (bottom) Ryann Ford Photography

p. 18: (top) Mark Lohman, design: Haefele Design; (bottom) Eric Roth, design: Leslie Fine Interiors, Boston

p. 19: Ryann Ford Photography, design: Poteet Architects

p. 20: Tria Giovan, design: Gregory Shano Interiors

p. 21: (top) Jim Westphalen Photography/Collinstock, design: Jennifer Palumbo Interior Designs; (bottom) Chris Luker Photography/ Collinstock, design: Nequette Architecture & Design

p. 22: (top) Mark Lohman; (bottom) © Brian Vanden Brink, design: Sally Weston Associates

p. 23: Hulya Kolabas

pp. 24–25: Helen Norman

p. 26: Mark Lohman, design: QualCraft Construction Inc.

p. 27: (top) Mark Lohman, design: Alison Kandler Interior Design; (bottom) Andrea Rugg Photography/ Collinstock, design: Hendel Homes

p. 28: (top) davidduncanlivingston. com; (bottom) Carolyn Bates

p. 29: (bottom) Susan Teare, design: Conner & Buck Design Build

CHAPTER 2

p. 30: davidduncanlivingston.com

p. 32: (top) Ryann Ford Photography; (bottom) Trent Bell Photography for Bloom Architecture

p. 33: © Brian Vanden Brink, design: Polhemus Savery DaSilva Architects

p. 34: Helen Norman, design: Lauren Liess

p. 35: (top) Trent Bell Photography for Ingunn Milla Joergensen; (bottom) Mark Lohman, design: Noelle Schoop Interior Design

p. 36: (top) Jim Westphalen Photography/Collinstock, design: Connor Homes; (bottom) Mark Lohman, design: Haefele Design

p. 37: Mark Lohman, design: Caroline Burke Designs

pp. 38–39: Susan Teare, design: Cecilia Redmond, architect: Haynes & Garthwaite; builder: Silver Maple Construction

pp. 40–41: Andrea Rugg Photography/ Collinstock, design: Lana Barbarossa, Allied ASID, designer

p. 42: (top) davidduncanlivingston. com; (bottom) Andrea Rugg Photography; design: Kate Roos Design

p. 43: (bottom) Helen Norman

pp. 44–45: Chipper Hatter, design: Kristianne Watts, KW Designs

p. 46: Mark Lohman, design: Noelle Schoop Interior Design

p. 47: (top) Mark Lohman, design: Alison Kandler Interior Design; (bottom) davidduncanlivingston.com

p. 48: Hulya Kolabas, design: Tiffany Eastman Interiors & Susan Glick Interiors

p. 49: (top) Ryann Ford Photography, design: Emily Seiders, Studio Seiders; (bottom left) davidduncanlivingston. com; (bottom right) Helen Norman, design: George Svatos; styled by Susan Victoria

p. 50: (top) Helen Norman; (bottom) Jim Westphalen Photography/ Collinstock, design: Jennifer Palumbo Interior Designs

p. 51: Ryann Ford Photography, design: Cravotta Interiors

p. 52: Helen Norman, architect: Ken Pursley; styled by Rebecca Omweg

p. 53: (top left) Tria Giovan, design: Kevin Spearman; (top right, bottom) Chipper Hatter, design: Corine Maggio, CM Natural Designs

CHAPTER 3

p. 54: Ryann Ford Photography

p. 56: Hulya Kolabas, design: Anna Burke Interiors

p. 57: (top) Ryann Ford Photography, design: Michael Deane Studio; (bottom) Jim Westphalen Photography/ Collinstock, design: Lillian August Design (architect: TruexCullins Architecture; builder: Roundtree Construction)

p. 58: (top) Rob Karosis Photography/ Collinstock, design: PKsurroundings; (bottom) Helen Norman

p. 59: (left) davidduncanlivingston.

com; (top right) Mark Lohman; (bottom right) Mark Lohman, design: Alison Kandler Interior Design

p. 60: Rob Karosis Photography/ Collinstock, design: C. Randolph Trainor Interiors, architect: Samyn-D'Elia Architects

p. 61: (top) davidduncanlivingston. com; (bottom) Mark Lohman

p. 62: Olson Photographic, design: Linda Lee Bauer, Classic Kitchens

p. 63: (top) Trent Bel Photography for Brian Beaudette; (bottom) Randy O'Rourke, design: Hudson Valley Preservation

p. 64: Eric Roth, design: Peter Breese Architects

p. 65: (top, bottom) Mark Lohman, design: Haefele Design

pp. 66–67: Mark Lohman, design: Alison Kandler Interior Design

p. 68: (top) Mark Lohman, design: Jeff Troyer Architect; (bottom) davidduncanlivingston.com

p. 70: Andrea Rugg Photography/ Collinstock, design: Bella Custom Homes

p. 71: (top left, top right) Andrea Rugg Photography/Collinstock, design: Kate Roos Design; (bottom) Susan Teare, design: Jillian Bartolo, Peregrine Design/Build

p. 72: Andrea Rugg Photography, design: J. Kurtz Design

p. 73: (top right, bottom) Mark Lohman, design: Alison Kandler Interior Design

p. 74: (top) Tria Giovan; (bottom) Helen Norman

p. 75: (left, right) © Brian Vanden Brink; design: Hutker Architects

p. 76: davidduncanlivingston.com

p. 77: (top right) Hulya Kolabas, design: Avo Construction; (bottom) Mark Lohman

p. 78: Rob Karosis Photography/ Collinstock, design: Knickerbocker Group

p. 79: (top left) Rob Karosis Photography/Collinstock, design: PKsurroundings; (top right) Susan Teare, design Mary Beth Stilwell; builder: Silver Maple Construction; (bottom) Jo-Anne Richards, design: Ines Hanl, The Sky is the Limit Design

p. 80: (left) Rob Karosis Photography/ Collinstock, design: C. Randolph

Trainor Interiors; (right) davidduncanlivingston.com
p. 81: (top) Emily Followill Photography/Collinstock, design: Peachtree Architects; (bottom) Andrea Rugg Photography/Collinstock, design: Otogawa-Anschel Design+Build
p. 82: (top) Chipper Hatter, design: Kristianne Watts, KW Designs; (bottom) Susan Teare, design: Cecilia Redmond, architect: Haynes & Garthwaite, builder: Silver Maple Construction
p. 83: (top left) Jim Westphalen Photography/Collinstock, design: Connor Homes; (top right, bottom right) Stacy Bass, design: Wende Cohen for Bungalow; (bottom left) Susan Teare, design: Classic Home, Charlotte, VT
p. 84: Susan Teare, design: Cushman Design Group, general contractor: Tom Herrington
p. 85: (top left) Andrea Rugg Photography/Collinstock, design: Anchor Builders; (top right) Mark Lohman, design: QualCraft Construction Inc.; (bottom) davidduncanlivingston.com
pp. 86-87: Helen Norman

CHAPTER 4

p. 88: Helen Norman, design: Lauren Liess
p. 90: Susan Teare, design: Cecilia Redmond, architect: Haynes & Garthwaite, builder: Silver Maple Construction
p. 91: (top) Olson Photographic, design: Sally Scott Interior Design; (bottom) Helen Norman, design: Lauren Liess
p. 92: (top) Chipper Hatter, design: Corine Maggio, CM Natural Designs; (middle, bottom), Andrea Rugg Photography, design: Kate Roos Design
p. 93: Mark Lohman, design: Alison Kandler Interior Design
p. 94: Mark Lohman, design: Nick Norris Cabinetry
p. 95: (top left, top right) Chipper Hatter, design: Cabinet Factory Outlet Plus; (bottom) Andrea Rugg Photography, design: MSR Design
p. 96: (top) Mark Lohman, design, Jeff Troyer Architect; (bottom) Jo-Anne

Richards, design: Ines Hanl, The Sky is the Limit Design
p. 97: (top) davidduncanlivingston. com; (bottom) Jo-Anne Richards, design: Rus Collins, Zebra Group
p. 98: Helen Norman
p. 99: (top) davidduncanlivingston. com; (bottom) Hulya Kolabas, design: Avo Construction
p. 100: Chipper Hatter, design: Kristianne Watts, KW Designs
p. 101: (top left) Mark Lohman, design: Haefele Design; (top right) Hulya Kolabas; (bottom) Mark Lohman, design: QualCraft Construction Inc.
p. 102: Mark Lohman
p. 103: (top) Trent Bell Photography, design: Elliott & Elliott Architecture; (bottom left) Mark Lohman, design: Alison Kandler Interior Design; (bottom right) Emily Followill Photography/Collinstock, design: Design Galleria Kitchen and Bath Studio
p. 104: courtesy GE Appliances
p. 105: (top) Mark Lohman, design: Haefele Design; (bottom left, bottom right) Olson Photographic, design: Peter Genovese, Putnam Kitchens
p. 106: Hulya Kolabas, design: Raquel Garcia Design
p. 107: (top left) Emily Followill Photography/Collinstock, design: Melanie Davis Design, architect: Historical Concepts; (top right) Tria Giovan, design: Phillip Sides Interior Design; (bottom) davidduncanlivingston.com
p. 108: (top) Chipper Hatter, design: Home Improvements Group; (bottom) Mark Lohman, design: Haefele Design
p. 109: (top) Ryann Ford Photography; (bottom) Andrea Rugg Photography; design: Kate Roos Design
p. 110: (top) Emily Followill Photography/Collinstock; design: Courtney Dickey; architect: T.S. Adams Studio, Architects; (bottom left, bottom right) Andrea Rugg Photography/Collinstock, design: Rosemary Merrill Design, Kate Roos Design
p. 111: Emily Followill Photography/ Collinstock, design: Carter Kay Interiors
p. 112: (top) Mark Lohman; (bottom)

Mark Lohman, design: Alison Kandler Interior Design
p. 113: (top) Andrea Rugg Photography/Collinstock, design: Anchor Builders; (bottom) Andrea Rugg Photography/Collinstock, design: Otogawa-Anschel Design+Build
p. 114: Mark Lohman
p. 115: (top) Emily Followill Photography/Collinstock, design: Design Galleria Kitchen and Bath Studio; (bottom) Donna Griffith Photography/Collinstock, stylist: Ann-Marie Favot
p. 116: (top) Mark Lohman, design: Noelle Schoop Interior Design; (bottom) Chipper Hatter, design: Kristianne Watts, KW Designs
p. 117: (top) Virginia Hamrick Photography/Collinstock, design: Karen Turner/KTK Design; (bottom) Ryann Ford Photography, design: Julie Dodson, Dodson Interiors
p. 118: (top) Susan Teare, design: Cecilia Redmond; architect: Haynes & Garthwaite; builder: Silver Maple Construction; (bottom left, bottom right) Andrea Rugg Photography/ Collinstock, design: David Heide Design Studio
p. 119: (left) Mark Lohman, design: Noelle Schoop Interior Design; (right) Chipper Hatter, design: Home Improvements Group
pp. 120–121: Mark Lohman, design: Noelle Schoop Interior Design

CHAPTER 5

p. 122: Ryann Ford Photography, design: Stone Textile Design Studio
p. 124: Jo-Ann Richards, design: Colleen Buker, Colleen Buker Design
p. 125: (top) Emily Followill Photography/Collinstock, design: Frank G Neely Design Associates, Wendy Meredith Interiors; (bottom left) Hulya Kolabas, design: Mar Silver Interiors; (bottom right) Andy Frame Photography/Collinstock
p. 126: (top) © Brian Vanden Brink; (bottom) Susan Teare, design: Elizabeth Herrmann Architecture + Design; builder: Red House
p. 127: (top) davidduncanlivingston. com; (bottom) Mark Lohman, design: Haefele Design
p. 128: Tria Giovan

p. 129: (top) Emily Followill Photography/Collinstock, design: Chenault James Interiors; (bottom left) Emily Followill Photography/ Collinstock, design: Carter Kay Interiors; (right) Subtle Light Studio/ Collinstock, design: City Builders
p. 130: (left) Andrea Rugg Photography/Collinstock, design: TreHus Architects + Interior Designers + Builders; (top right) Andrea Rugg Photography/Collinstock, design: U+B Architecture & Design; (bottom right) Mark Lohman Photography/ Collinstock
p. 131: (bottom left) Andrea Rugg Photography/Collinstock, design: Sicora Design/Build; (bottom right) Subtle Light Studio/Collinstock
p. 132: Eric Roth, design: Butz + Klug Architecture
p. 133: (top right) Andrea Rugg Photography/Collinstock, design: Otogawa-Anschel Design+Build; (bottom) Scott Dorrance Photography/ Collinstock
p. 134: Tria Giovan
p. 135: (top right) Chipper Hatter, design: Roomscapes; (bottom left) Chris Luker Photography/Collinstock, design: Adams Cerndt Design Group; (bottom right) Andrea Rugg Photography/Collinstock, design: Kate Roos Design
p. 136: (top) Andrea Rugg Photography/Collinstock, design: Jacqueline Fortier Interior Designer; (bottom) Andrea Rugg Photography, design: MSR Design
p. 137: (bottom left) Jim Westphalen Photography/Collinstock, design: Birdseye Design/Build; (right) Andrea Rugg Photography, design: CF Design
p. 138: (top, middle, bottom) Mark Lohman
p. 139: Susan Teare, design: Brown + Davis Design, Architects, builder: Conner & Buck Builders, cabinets: Mark Simon
p. 140: (top) Trent Bell Photography for GO Logic; (bottom) Ryann Ford Photography, design: Kelly Moseley, Anabel Interiors
p. 141: (top right) Helen Norman, design: Lauren Liess; (bottom right) davidduncanlivingston.com
p. 142: (top) Andrea Rugg

Photography, design: Kate Roos Design; (bottom) Stacy Bass, design: Angela Camarda, Lillian August

p. 143: (top right) Olson Photographic, design: Joe Currie, Capitol Design; (bottom) Emily Followill Photography/Collinstock, design: Carter Kay Interiors

p. 144: (top) Tria Giovan; (bottom) Susan Teare, design: Cushman Design Group; general contractor: Steel Construction

p. 145: (top) Emily Followill Photography/Collinstock, design: The Design Atelier, architect: Greg Busch Architects; (bottom) Mark Lohman, design: Haefele Design

p. 146: (left) Susan Teare, design: Hart Associates Architects, general contractor: Gilman Guidelli and Bellow, cabinets: Leicht; (top right) Mark Lohman, design: Noelle Schoop Interior Design; (bottom right) Chipper Hatter, design: Kristianne Watts, KW Designs

p. 147: (top) Hulya Kolabas, design: Think Chic Interiors; (bottom) Eric Roth, design: Dalia Kitchen Design

p. 149: (top) davidduncanlivingston. com; (bottom) Chipper Hatter, design: Corine Maggio, CM Natural Designs

CHAPTER 6

p. 150: Helen Norman, design: Lauren Liess

p. 152: (top) davidduncanlivingston. com; (bottom) Andrea Rugg Photography/Collinstock, design: Liz Schupanitz Designs

p. 153: (top right) Emily Followill Photography/Collinstock, design: Brian Patrick Flynn, Flynnside Out Productions; (bottom) courtesy Kohler

p. 154: (left) davidduncanlivingston. com; (right) Hulya Kolabas

p. 155: Rob Karosis Photography/Collinstock, design: Smith & Vansant Architects

p. 156: (left) Mark Lohman, design: Alison Kandler Interior Design; (right) Ryann Ford Photography, design: Robin Colton Studio

p. 157: (top right) Ryann Ford Photography, design: Amity Worrel & Co.; (middle right) Rob Karosis Photography/Collinstock, design: C. Randolph Trainor Interiors; (bottom right) Matthew Quinn, Home Refinements

p. 158: (top) Andrea Rugg Photography/Collinstock, design:

Hendel Homes; (bottom) Tria Giovan, design: Babcock Peffer Design

p. 159: (top) © Brian Vanden Brink, design: Sally Weston Associates; (bottom) Tria Giovan, design: Greg Shano

p. 160: (top) courtesy Kohler; (bottom) Susan Teare, design: Cushman Design Group, general contractor: Gregory Construction

p. 161: (top) courtesy Kohler; (bottom) Eric Roth, builder: Payne Bouchier Fine Builders

p. 162: Ryann Ford Photography

p. 163: (top left) Mark Lohman; (top right) Ryann Ford Photography; (bottom left) Eric Roth, design: Decori Design, Boston; (bottom right) © Brian Vanden Brink, design: Polhemus Savery DaSilva Architects

p. 164: Stacy Bass, design: Dovecote

p. 165: (top left) Jo-Anne Richards, design: Rus Collins, Zebra Group; (top right) courtesy Kohler; (bottom) Jo-Anne Richards, design: Ines Hanl, The Sky is the Limit Design

p. 166: Mark Lohman, design: Alison Kandler Interior Design

p. 167: (top left) Helen Norman; (top right) Ryann Ford Photography; (middle) Olson Photographic, design: Sally Scott Interior Design; (bottom) Emily Followill Photography/Collinstock; design: Melanie Turner Interiors, builder: Benecki Fine Homes

p. 168: (top) Olson Photographic, design: Hedy Shashaani, Jack Rosen Custom Kitchens; (bottom) Mark Lohman, design: Haefele Design

p. 169: (top, bottom) courtesy Kohler

p. 170: davidduncanlivingston.com

p. 171: (top, middle left, bottom) courtesy Kohler; (middle right) davidduncanlivingston.com

CHAPTER 7

p. 172: Hulya Kolabas

p. 174: (top) Andrea Rugg Photography/Collinstock, design: Chester-Hoffman & Associates; (bottom) Mark Lohman, design: Noelle Schoop Interior Design

p. 175: (top right) Emily Followill Photography/Collinstock, design: Liz Williams Interiors; architect: D. Stanley Dixon Architect; (bottom) Andrea Rugg Photography, design: Quartersawn Design Build

p. 176: (top) © Brian Vanden Brink, design: Hutker Architects; (bottom)

Susan Teare, design: Brian Hamor, Hamor Architecture Associates; builder: Sean Gyllenborg, Gyllenborg Construction

p. 177: (left) Helen Norman; (right) Emily Followill Photography/Collinstock; design: Melanie Turner Interiors, builder: Benecki Fine Homes

p. 178: Hulya Kolabas, design: Mar Silver Interiors

p. 179: (top) Paul Dyer Photography/Collinstock, design: Feldman Architecture; (bottom) Virginia Hamrick Photography/Collinstock, design: W.A. Marks Fine Woodworking

p. 180: Eric Roth, design: Mark Christofi Interiors

p. 181: (top right) Tria Giovan, design: John Bjørnen, Bjørnen Design; (bottom) Tria Giovan, design: Ken Gemes Interiors

p. 182: (top) Susan Teare, design: Silver Maple Construction; (bottom) Mark Lohman, design: Leah Anderson

p. 183: (top right) davidduncan livingston.com; (bottom) Andrea Rugg Photography/Collinstock, design: Hendel Homes

p. 184: Susan Teare, design: Stephen Wanta Architect, builder: Peregrine Design/Build

p. 185: (top) Andrea Rugg Photography, design: Lawrence Rugg Architecture and De La Cruz Interior Architecture + Design; (bottom left) Jo-Ann Richards, design: Colleen Buker, Colleen Buker Design; (bottom right) Chipper Hatter, design: Home Improvements Group

p. 186: Ryann Ford Photography, design: Redbud Custom Homes

p. 187: (top) © Brian Vanden Brink, design: Hutker Architects; (bottom) Trent Bell Photography for GO Logic

p. 188: (top) Carolyn Bates, design: Ali White, Kitchens by Design; (bottom) davidduncanlivingston.com

p. 189: (top) Mark Lohman, design: Jeff Troyer Architect; (bottom) Mark Lohman, design: Haefele Design

CHAPTER 8

p. 190: Tria Giovan

p. 192: (top) Helen Norman, design: Lauren Liess; (bottom) Stacy Bass, design: Angela Camarda, Lillian August

p. 193: (top right) Chipper Hatter, design: Habify; (bottom right) Susan Teare, design: Brian Hamor, Hamor

Architecture Associates, builder: Scott Driscoll, Home Tech Construction

p. 194: (top) Chipper Hatter, design: Countertop Shoppe; (bottom) Hulya Kolabas, design: Raquel Garcia Design

p. 195: (top, bottom) davidduncan livingston.com

p. 196: (top) Helen Norman; (bottom) Tria Giovan

p. 197: (top right) Eric Roth, design: Oak Hill Architects; (bottom right) Tria Giovan; design, Meg Lonergan

p. 198: (top) Stacy Bass, design: Shelley Morris Interiors; (bottom) Tria Giovan

p. 199: (top) Tria Giovan; (bottom right) Ryann Ford Photography

p. 200: (top) © Brian Vanden Brink; design: Eck MacNeely Architects; builder: Wright Ryan Builders; (bottom left) Emily Followill Photography/Collinstock, design: T.S. Adams Studio, Architects, builder: Kenneth P. Dooley Construction; (bottom right) © Brian Vanden Brink, design: Hutker Architects

p. 201: Tria Giovan, design: Kevin Spearman

p. 202: © Brian Vanden Brink, design: Hutker Architects

p. 203: (top left) Olson Photographic, design: Callaway Wyeth Architects; (top right) © Brian Vanden Brink; (bottom) Mark Lohman, design: Cynthia Marks Design

p. 204: (top) Eric Roth, design: Matthew Sapera Fine Homes; (bottom) Tria Giovan, design: Scott Sanders

p. 205: (top right) Tria Giovan, design: Lynn Morgan; (bottom) Hulya Kolabas, design: Raquel Garcia Design

p. 206: (top, bottom) Mark Lohman, design: Alison Kandler Interior Design

p. 207: (top) © Brian Vanden Brink, design: Hutker Architects; (bottom) davidduncanlivingston.com

p. 208: (top) Tria Giovan, design: Kate Jackson; (bottom) Susan Teare, design: Jennifer Lane Architecture & Design, builder/cabinetry: Cypress Woodworks

p. 209: (top right) davidduncan livingston.com; (bottom right) Eric Roth

p. 210: (top, bottom) Helen Norman, design: Lauren Liess

p. 211: (top right) Mark Lohman, design: Alison Kandler Interior Design; (bottom) Tria Giovan, design: Phoebe Howard

INDEX